The Blue Book on China's Ecological Red Lines 2023

China Land Surveying and Planning Institute

The Blue Book on China's Ecological Red Lines 2023

China Land Press

Springer

China Land Surveying and Planning Institute
Beijing, China

ISBN 978-981-95-0050-5 ISBN 978-981-95-0051-2 (eBook)
https://doi.org/10.1007/978-981-95-0051-2

Jointly published with China Land Press Ltd.
The print edition is not for sale in China (Mainland). Customers from China (Mainland) please order the print book from: China Land Press Ltd.
ISBN of the Co-Publisher's edition: 978-7-5200-1161-7

This Springer imprint is published by the registered company Springer Nature Singapore Pte Ltd.
The registered company address is: 152 Beach Road, #21-01/04 Gateway East, Singapore 189721, Singapore

Foreword

Ecological Civilization marks a new era of human civilization, representing an introspection on past patterns of development and serving as an inevitable choice for coping with global climate change and maintaining the sustainable development of the life system on Earth. China has carried out a series of pioneering works in ecological conservation and restoration, making the construction of an Ecological Civilization and a Beautiful China a significant step forward.

The Ecological Red Lines (ERLs) serve as the bottom line and lifeline for safeguarding and maintaining territorial ecological security, and they represent a significant institutional innovation in the construction of an Ecological Civilization. Delineating and strictly adhering to ERLs is a significant policy decision and deployment in the construction of Ecological Civilization in the new era. It is also an important approach to addressing ecological environmental issues and supporting high-quality development with a high-quality ecological environment. As a pioneering innovation, scientifically delineating ERLs presents significant challenges. This difficulty arises not only from the complex task of identifying key ecological functions within these areas and delineating these zones as completely as possible, but also from the need to balance ecological security with developmental needs and to deal with various contradictions and conflicts among demands for different land uses. It also has to define the ERLs on the ground, so as to achieve consistency in "data, line and map". For this purpose, the Ministry of Natural Resources along with the Ministry of Ecology and Environment, the National Forestry and Grassland Administration and other Ministries coordinated with local governments at all levels to carry out this mission. Using the findings of the Third National Land Survey as the base map and data, and based on the evaluation of resources and environmental carrying capacity, the territorial development suitability, regulations were formulated successively to properly address various spatial contradictions and conflicts. The land and marine ERLs were delineated completely. Achievement was made in managing key ecological areas within one red line.

The total area covered nationwide by the ERLs is up to 3.19 million km^2, of which the land ecological protection red line accounts for more than 30% of the land area, including the protected areas, extremely important ecological function areas, extremely fragile ecological areas and areas with potential important ecological value. It has laid the foundation for the construction of a territorial ecological security pattern and provided Chinese solutions for global ecological conservation.

As the Chairman of the National Expert Group on Delineation of ERLs, I have witnessed the hardships of delineation, and felt delighted to see that such a great mission has been successfully accomplished. The delineation of China's ERLs has been a focus of considerable concern for experts, scholars, governmental sectors and the public both domestically and globally. The outcome of the delineation provides strong support for strengthening ecological conservation and restoration and promoting Ecological Civilization construction. Compiled by the China Land Surveying and Planning Institute (CLSPI), this book systematically summarizes the process, methodology, achievements and practical cases of delineation of ERLs, and presents ideas and suggestions on strengthening the supervision and administration of ERLs and improving their system, which can help various sectors of society to better understand ERLs, and to foster a conscious awareness and broad consensus for ecological conservation.

The delineation of ERLs is not only a practice in China on ecological conservation and governance, but also an innovative solution for global ecosystem and biodiversity conservation. I look forward to the future where relevant government departments and the public work together to improve the supervision and administration of ERLs, strengthen the conservation and restoration of ecosystems, and leave behind a beautiful ecological space and a wonderful homeland for future generations.

Chinese Academy of Sciences
Beijing, China
August 2, 2023

Bojie Fu

Editorial Board

Preface

The Blue Book on China's Ecological Red Lines 2023, compiled by the China Land Surveying and Planning Institute (CLSPI), is the first blue book on the research outcomes of Ecological Red Lines (ERLs). This book systematically summarizes the process, methodology, achievements and practical cases of the delineation of ERLs. It offers ideas and suggestions for enhancing the supervision and administration of ERLs, as well as for establishing a robust long-term monitoring, assessment and early warning mechanism. This book will serve as a reference for policymakers and all sectors of society.

There are five chapters in the book. Chapter 1 is the introduction to the international and domestic background and the objectives of ERLs delineation in China. Chapter 2 is about progress and methodology. It systematically summarizes the working progress and mechanism of ERLs delineation, and the solutions for relevant conflicts. Chapter 3 is about achievements. It presents the general situation, spatial distribution, structure of land use, important ecological functions and typical ecological classes of ERLs. Chapter 4 is about management and control measures. It puts forward suggestions on strengthening the monitoring, assessment and early warning of ERLs and improving the long-term supervision mechanism based on investigation of the existing control protocols. Chapter 5 is about the practical cases of local governments. It presents the local practical experiences and ideas in delineating and controlling ERLs in nine provinces, namely, Jiangxi, Hainan, Heilongjiang, Zhejiang, Anhui, Guangdong, Sichuan, Qinghai and Fujian.

The book is the quintessence of collective wisdom. The editorial board mainly consists of scholars from relevant technical departments of CLSPI, who specialize in the study of ERLs and have long been involved in this research. The findings in this book draw on the valuable experience of institutions and research institutes engaged in the delineation of ERLs.

Our sincere thanks go to Nanjing Institute of Environmental Sciences of the Ministry of Ecology and Environment, Survey, Planning and Design Institute of National Forestry and Grassland Administration, Research Center for Eco-Environmental Sciences of Chinese Academy of Sciences, Information Center of the Ministry of Natural Resources, National Marine Data and Information Service,

Land Satellite Remote Sensing Application Center of the Ministry of Natural Resources, Strategy Research Center for Oil and Gas Resources of the Ministry of Natural Resources, Development Research Center of the China Geological Survey, Chinese Research Academy of Environmental Sciences, Chinese Academy of Environmental Planning of the Ministry of Ecology and Environment, Satellite Application Center for Ecology and Environment of the Ministry of Ecology and Environment, National Marine Environmental Monitoring Center of the Ministry of Ecology and Environment, Department of International Cooperation of the Ministry of Natural Resources, Bureau of Territorial and Spatial Planning of the Ministry of Natural Resources, Division of Map Affairs Administration of the Geo-information Management Department.

Special thanks are also given to Prof. Wang Qingqin of China Academy of Building Research, Prof. Peng Jian of Peking University, Prof. Liu Xuehua of Tsinghua University, Prof. Li Junsheng of the Chinese Research Academy of Environmental Sciences, Prof. Niu Shuli of the Chinese Academy of Sciences, Prof. Lyu Cai of Beijing Forestry University and Ms. Wang Hui of *China Daily* for their contributions to the English version of this book.

This book may contain errors due to time constraints and the authors' limitations; we welcome any corrections and feedback from our readers.

Beijing, China — China Land Surveying and Planning Institute
August 1, 2023

Competing Interests The author has no competing interests to declare that are relevant to the content of this manuscript.

The original version of the book has been revised. A correction to the book can be found at https://doi.org/10.1007/978-981-95-0051-2_6

Contents

Chapter 1
Introduction

The Ecological Red Lines (ERLs) refer to the protected areas, extremely important ecological function areas, extremely fragile ecological areas and areas with potential important ecological value, which must be conserved strictly and compulsorily. It is also the bottom line and lifeline of safeguarding and maintaining territorial ecological security. Delineation and strictly safeguarding ERLs is an important measure taken by the Chinese government in the new era.

1.1 Background

1.1.1 International Background

1.1.1.1 Global Climate Governance Calls for Action

Global warming has led to accelerated melting of glaciers, intensified desertification, rising sea levels and other issues. In recent years, the frequent occurrence of extreme weather has posed serious challenges to human survival and development. According to the State of the *Global Climate* 2022 issued by the World Meteorological Organization (WMO), 2015–2022 are likely to be the eight warmest years on record since 1850. During 2013–2022, the average rise of global sea level was about 4.62 mm/year. According to the report from the National Snow and Ice Data Center in the U.S., the Antarctic Ocean ice area shrank to 1.91 million km^2 on February 13, 2023, marking the lowest level on record. Therefore, improving climate governance has become a global consensus.

China is moving forward to accelerate the construction of a harmonious ecological system that respects nature and promotes green development. Since territorial space is the carrier of Ecological Civilization construction, the Chinese government has taken the initiative to seek "Nature-based Solutions" (NbS) in active response to climate change. By scientifically delineating ERLs, China aims to strictly conserve

China Land Surveying and Planning Institute, *The Blue Book on China's Ecological Red Lines 2023*, https://doi.org/10.1007/978-981-95-0051-2_1

and systematically restore forests, grasslands, wetlands and other ecosystems, so as to improve ecological functions and contribute to global climate action.

1.1.1.2 Biodiversity Conservation Needs National Solutions

Biodiversity is the basis for human survival and development, and it has a direct bearing on our well-being. At present, the rate of the global species extinction is accelerating. The loss of biodiversity and the degradation of the ecosystem pose a major risk to human survival and development. About a quarter of the world's species are currently at risk, with the rate of species extinction occurring at a pace 100 to 1000 times faster than the natural extinction rate.

China proposes to delineate and strictly manage ERLs, which identify key natural ecosystems, natural landscapes and natural relics into protected areas as the core components of ERLs. China emphasizes the integrity and connectivity of ecosystems and prioritizes the inclusion of regions critical for biodiversity conservation within the ERLs. This has led to extensive and holistic protection of vital ecological spaces, rare and endangered species, and their habitats, creating a distinctive Chinese approach to biodiversity conservation.

1.1.1.3 Sustainable Development Goals Welcome Forerunner

At present, global ecological crises, such as climate change, biodiversity loss and land degradation could be harmful to the long-term survival and sustainable development of mankind. Historical reviews show that many countries, including some developed countries, seek economic growth at the expense of the ecological environment. Once the ecological environment is damaged, greater price is needed to recover and govern it. Exploiting and even destroying nature without restraint will inevitably result in nature's retribution.

The Chinese government regards Ecological Civilization construction as the fundamental strategy for sustainable development, adhering to the course for green, inclusive and sustainable development. The government delineates and abides by ERLs, carries out large-scale afforestation, and strengthens the comprehensive prevention and control of desertification. China has been exploring a sustainable development path that gives priority to ecological, green and low-carbon development, and has become an important participant and contributor in global ecological conservation.

1.1.2 Domestic Background

1. China's ecosystems are inherently fragile and structurally imbalanced, with ongoing pressures not yet fundamentally alleviated.

First, China's ecosystems are inherently fragile. Ecologically fragile areas and the most ecologically fragile areas account for about 48% of China's land area. There are prominent ecological problems, such as desertification and salinization of land in the northwest and north of China, land loss and rocky desertification in the southwest and thinning of black soil in the northeast. The foundation of ecological security is not stable. Over a period of time, due to overexploitation and unregulated development of resources, the natural ecosystem has been displaced and destroyed. China faces increasingly grave problems including environmental tightening, resource constraints and ecological degradation.

Second, the structural imbalance of China's ecosystem is increasingly prominent. According to the results of the Third National Land Survey, the glacier area on the Qinghai-Xizang Plateau has retreated by 10% in the last 10 years, and the grassland in the northern sand prevention belt and the forest and wetland in the northeast forest belt have been reduced too much; excessive agricultural reclamation and grazing have caused imbalance of some grassland ecosystems, and the average overloading rate of livestock in key natural grasslands has reached 10%; the water conservation function of water source area is unstable, and the utilization rate of water resources in the Huang He, Hai He, Liao He and other basins is relatively high, causing insufficient ecological water volume; some important ecological security spaces have been eroded, and the habitats of important species have been fragmented. The ecological functions in some areas have degraded, and the descending trend in biodiversity has not been effectively restrained.

Third, ecological conservation in China is still under multiple pressures, and faces a lot of difficulties. The ecological environment quality of China has been improved constantly through unremitting efforts. However, China's Ecological Civilization construction still faces many contradictions and challenges. The foundation for the steady improvement of the ecological environment is not solid. The turning point from quantitative change to qualitative change has not arrived, while ecological security and sustainable development still face great threats. As China's economic and social progress has entered a phase of ever-accelerating green, low-carbon and high-quality development, it is necessary to provide more high-quality ecological products to meet people's needs for a better life. Meanwhile, it is necessary to have the capability to address prominent ecological and environmental issues, constantly strengthen ecosystem conservation, implement the strictest ecological and environmental conservation system to ensure that the natural environment is not threatened or damaged, thus improving the service functions of the ecosystem, reserving space for sustainable development, and leaving a beautiful home for future generations.

2. The spatial pattern and conservation mode which support harmonious coexistence between humans and nature need to be formed urgently.

   To further conserve and optimize ecological space, China has implemented plans for national ecological function zones, main functional zones, strategies and action plans for biodiversity conservation, etc. China has also expedited the establishment of protected areas which has played an important role in ecological protection. However, challenges in ERLs delineation and conservation effectiveness persist, detailed as follows:

First, the supply of ecological space is difficult to meet people's needs for a better life. Previously, China's ecological security was structured around the "two barriers and three belts" pattern, with the Qinghai-Xizang Plateau ecological barrier, the Loess Plateau-Sichuan-Yunnan ecological barrier, the northeast forest belt, the northern sand prevention belt, and the southern hilly mountain belt serving as its backbone. This pattern and various protected areas were concentrated west of the Heihe-Tengchong Line, primarily in the central and western regions of China. Insufficient consideration is given to the densely populated eastern regions. As a result, there is a need to enhance the ecological security pattern and spatial distribution. The implementation of national and regional ecological security strategies is hindered by an absence of systematic ecological conservation planning across macro, meso, and micro scales.

Second, the integrity and connectivity of the ecosystem is insufficient. Currently, a cohesive approach to the overall protection, systematic restoration, and integrated management of different ecosystems—such as land, freshwater, and marine—is yet to be established. The problems such as overlapping and poor connectivity have existed among the established conservation areas. Some natural ecosystems, and rare and endangered species have not been conserved effectively.

Third, there is an absence of contradiction coordination regulations for development and conservation. The ecological conservation areas specified in various plans and actions have reached more than half of the national territory before. Coexistence of development and conservation occurs in a large range of important ecological areas. There are obvious space contradictions in urban and rural construction, mineral exploitation, resource utilization and ecological conservation. The lack of clear and detailed control measures and space access rules makes it difficult to implement strict conservation. Key ecological areas face the threat of gradual encroachment by development and construction activities.

3. It is necessary to establish a unified, coordinated and full-cycle ecological conservation management system.

Previously, China's ecological conservation management functions were scattered across various departments, based on the division of labor by resource elements, which meant a lack of a powerful, unified and efficient ecological conservation management system.

First, the unified ecological conservation system should be improved. In the past, key ecological conservation areas and resource elements such as nature reserves, tourist attractions, geological parks, drinking water sources, forests, grasslands and wetlands were managed by different departments. A series of relevant laws, regulations and management provisions were issued successively, laying a solid foundation for building a comprehensive ecological conservation management system. However, these areas have wide spatial coverage and complex internal structures, which are related to the coordination of multiple dimensions such as ecological conservation, regional development and people's well-being. The management of these areas implicates various levels and entities, such as central and local governments, different government departments, and both governmental and individual stakeholders. It is necessary to explore a set of unified control systems that are integrated and regional, scientific and feasible.

Second, an ecological conservation supporting system for the whole process at multi-level is needed. Due to the vastness of the ecological area, the fragility of the ecosystem, the complexity of human activities and the emergency of some ecological destruction events, periodical monitoring and evaluation cannot fulfill the requirements of accurate and efficient management, while daily monitoring and timely warning system has not yet formed. It is necessary to establish a multi-level evaluation technology system of monitoring and conservation effectiveness, which covers all major ecological areas and country-province-municipality-county-town-parcel in scale. Third, the public participation mechanism is imperfect. There is a lack of public participation channels. The awareness of respecting nature, accommodating and conserving nature has not been firmly established. China needs to strengthen the information disclosure of the spatial distribution, conservation effectiveness and monitoring of important ecological areas. The government also needs to guide the public participating in daily supervision, fully leveraging the public's proactive role as the subject of rights and responsibilities in ecological conservation.

1.2 Objectives of ERLs Delineation

1.2.1 Strictly Holding the Bottom Line of Territorial Ecological Security

The ERLs are the bottom line and lifeline of ensuring and maintaining territorial ecological security. The delineation of ERLs should be based on the physical geographical pattern and consider the authenticity, integrity and systematicity of natural ecosystems. Priority should be given to essential ecological functional areas, such as water conservation, biodiversity conservation, soil conservation, windbreak and sand fixation and coastal conservation. The most ecologically fragile areas, such as those prone to soil erosion, desertification, rocky desertification and coastal erosion and the natural protected areas for conserving typical natural ecosystems, rare and endangered species of wildlife and natural sites of special significance, together with other areas with potentially important ecological value, should also be incorporated into the ERLs.

In addition, considering the importance of maintaining national and regional ecological security, typical ecosystems such as wetlands, habitats of species with very small populations, mangroves, glaciers and permanent snow, coral reefs and seagrass meadows should be covered within the ERLs, thus improving the diversity, stability and continuity of ecosystems, defending the boundary of natural ecological security and strengthening the foundation of territorial ecological security.

1.2.2 Optimizing the Pattern of Territorial Ecological Security

On the basis of fully understanding China's physical geographical condition, the current status and evolution laws of ecosystems, the challenges to achieve the territorial ecological security pattern are following:

1. The needs for ecosystem conservation at different scales range from the world, country to the area.
2. The differences in natural ecosystem characteristics and differences in economic and social development between the east and west sides of the "Heihe-Tengchong Line".
3. Enhancing the distribution of ecological functional areas in densely populated urban areas in southeast China, improving the ability to supply ecological security and ecological products to key regions such as the Beijing-Tianjin-Hebei Region, the Chang Jiang Economic Zone and the Guangdong-Hong Kong-Macao Greater Bay Area.
4. Strengthening the territorial ecological security framework and cross-regional ecological connectivity, integrating the "three zones and four belts", namely, the Qinghai-Xizang Plateau ecological zone, the Huang He key ecological zone, the Chang Jiang key ecological zone, the northeast forest belt, the northern sand prevention belt, the southern hilly mountain belt, and the coastal belt, across three major geographical and climate zones of the country.
5. Making the coastal zone and other trans-regions play the role of ecological corridor, and smoothing the ecological connection between plateaus, important mountains, major plains and oceans.

By solving the above challenges, a consolidated national barrier will have been significantly reinforced, thereby ensuring territorial ecological security.

1.2.3 Promoting the Innovation of the Ecological Civilization System

The delineation of ERLs is a landmark achievement and a major institutional innovation of the Ecological Civilization construction in China. The meanings are:

1. It improved the administration system. According to the integrity and systematicity of the ecosystem and its internal rules, the situation of overlapping management has to be changed. The use control should be extended to all naturally ecological environments, with one single department, namely, MNR, taking responsibility for the management of land use across the national territory.
2. It expanded the adoption of the control system for national territory, implemented integrated conservation and unified management of mountains, rivers, forests, farmlands, lakes, grasslands and sands.
3. It linked the ERLs with the laws and regulations of protected areas, to improve the management and control rules for the ERLs according to the actual conditions of different regions, zoning functions and conservation objects, thus advancing the scientificity and practicality of the ERLs.
4. It provided support to improve the national ecological conservation compensation system, to explore the "Two Mountains" transformation and the path to realizing the value of ecological products and to promote the establishment of a horizontal ecological compensation mechanism.
5. It built a multi-level monitoring and conservation effectiveness evaluation system, covering the ERLs and surrounding areas to enhance monitoring and supervision capability and support high-quality development with high-level conservation, supported by the "space-air-ground-network" natural resources survey and monitoring system, and "one map" of national spatial planning.

Chapter 2
Progress and Methodology

2.1 Delineation Progress

The delineation of the ERLs and abiding strictly is one of the strategic deployments made by the Chinese government. As one of the main departments responsible for Ecological Civilization construction, the Ministry of Natural Resources (MNR), together with the Ministry of Ecology and Environment (MEE), the National Forestry and Grassland Administration (NFGA) and other departments, jointly carried out and fully completed the ERLs delineation, covering both the land and the marine. In view of the time sequence, the process can be divided into two parts—before and after the institutional reform in 2018; in view of the course, it can be divided into two stages—formal proposal and practical exploration, optimization and improvement.

2.1.1 Phase 1: Proposal and Practical Exploration

In October 2011, the Chinese government officially put forward the concept of "Ecological Red Line" and proposed to carry out the delineation of the ERLs. Therefore, the ERLs delineation in various regions has officially entered a stage of rapid progress. In March 2012, the Ministry of Environmental Protection deployed and carried out pilot work on the ERLs delineation in Inner Mongolia, Jiangxi, Guangxi and Hubei, etc. In addition, the exploration of the ERLs system and ecological compensation mechanism was conducted in Shanxi, Shandong, Tianjin, Beijing, Guizhou, Jiangsu, Shanghai, and Xinjiang, etc. During the same period, to strengthen marine ecological conservation, the State Oceanic Administration promoted the delineation of marine ERLs of Bohai Sea.

In January 2017, the Chinese government planned that the delineation of the ERLs should be completed by the end of 2020. The Ministry of Environmental

China Land Surveying and Planning Institute, *The Blue Book on China's Ecological Red Lines 2023*, https://doi.org/10.1007/978-981-95-0051-2_2

Protection and the National Development and Reform Commission formulated and issued technical guidelines for delineation, in which clarified the principles, working procedures, technical processes and requirements for results.

The Ministry of Environmental Protection took the lead to complete the delineation of the ERLs in Beijing-Tianjin-Hebei Region, provinces along the ChangJiang Economic Belt, Ningxia and other 15 provinces (autonomous regions and municipalities directly under the central government) in February 2018 and it was issued and implemented by the provincial governments. In October 2018, the technical review of the ERLs in 16 provinces (autonomous regions) including Shanxi was completed. In December 2017, the State Oceanic Administration took the lead in completing the delineating, issuing and implementing of marine ERLs.

2.1.2 Phase 2: Optimization and Improvement

In 2018, the Chinese government carried out institutional reform of the State Council. The MNR assumed the responsibility of establishing, implementing and supervising the territorial and spatial planning system, including the delineation and supervision of the ERLs. In order to develop this project, the MNR organized many times field surveys and forums to hear the opinions of relevant departments, local governments and experts. The MNR seriously concerned about the prominent contradictions reported by the local governments and found that there were many conflicts among the ERLs and permanent basic farmlands, towns and villages, mining rights and artificial commercial forests due to the inaccurate basic data and unclear boundaries of original natural protected areas.

In order to resolve the contradictions and conflicts among three controlled lines, namely the ERLs, permanent basic farmland protect lines, urban development boundary, considering the overall development and security, starting from July 2019, the MNR, MEE and NFGA have jointly carried out the evaluation and adjustment of the ERLs in 15 provinces (autonomous regions and municipalities directly under the central government) and the delineation of the ERLs in 16 provinces (autonomous regions) approved by the State Council on the basis of existing work.

According to the principles of scientificity, simplicity and operability, MNR compiled the technical guideline, clarifying the coordination rules for the boundary contradictions of three controlled lines. At the same time, according to the Central Government's guideline on establishing a system of protected areas with national parks as the subject, the MNR and NFGA jointly issued a notice on the preliminary work related to optimizing the scope and functional zoning of nature reserves. They have planned and carried out the integration and optimization of protected areas, and put the optimized protected areas into the ERLs.

The evaluation adjustment and delineation of the ERLs covered 31 provinces (autonomous regions and municipalities directly under the central government). The main scope of work is 9.6 million km^2 in land area and 320,000 km^2 in the nearshore sea areas. The marine protected areas outside the nearshore sea areas have been included in the marine ERLs.

Through the joint efforts of different departments at all levels, and after multiple rounds of basic data connection, conflict analysis, layout optimization, adjustment with full coverage, combined with the *Outline of the National Territorial and Spatial Planning* (2021–2035) (hereinafter referred to as the National Outline), the nationwide ERLs were completed for the first time and approved by the Central Committee of the Communist of China (CPC) and the State Council. All the ERLs are delineated on the planning map and recorded in the database and land on each parcel. The ERLs are also included in the "one map" supervision system to realize digital management, serving as the basis for the approval of land and marine use.

2.2 Delineation Methodology

2.2.1 Basic Principles

Based on the results of the Third National Land Survey, according to the results of the evaluation of resources and environmental carrying capacity, the territorial development suitability (hereinafter referred to as the "Dual Evaluation"), as well as the principles of strengthening overall planning, the combination of delineation and management, conservation priority and seeking truth from facts, China is aiming to coordinate various spatial conflicts in the "one map" of territorial and spatial planning to complete the delineation of the ERLs across the country.

1. Emphasizing on overall planning. Once the local government delineated three controlled lines according to the National Outline and integrate "Three Zones and Three Lines", namely the ecological space, agricultural space, urban space, and the ERLs, permanent basic farmland protect lines, urban development boundary, it should incorporate them into local territorial and spatial plans for strict implementation. The government is responsible for promoting the delineation of the ERLs and the integration and optimization of protected areas. All such areas should be brought into the ERLs, in addition to the central areas of tourist attractions.
2. Combination of delineation and administration. Administration has been considered since the start of delineation. The clear principle of conflict coordination was set up from the beginning, the Chinese government formulated and issued more than 10 documents and nearly 100 rules and clarified the specific requirements for handling conflicts. To fully connect with the conflict management rules at the time of delineation (Table 2.1), in August 2022, the MNR, MEE and NFGA jointly issued guidelines to strengthen the management of the ERLs and clarify the subsequent management requirements.
3. Giving priority to conservation. On the one hand, the Chinese government adheres to the priority of ecology, and delineates areas based on the physical

Table 2.1 Protocols for conflict management of ERLs delineation

Conflicts type	Management rules
Permanent basic farmlands	1. In the core reserve of protected areas, the original permanent basic farmlands are no longer managed as before, but kept in the ERLs as common farmlands and can be converted into ecological land gradually and orderly in the future according to the policy of ecological reclamation
	2. Outside the core reserve of protected areas, the concentrated contiguous permanent basic farmlands and long-term stable farmlands can be transferred out of protected areas and the ERLs. While the fragmented or scattered farmlands can be retained within the ERLs as common farmland
	3. Permanent basic farmland and long-term stable farmland within the following will never be adjusted out:
	Five national parks: Three-River-Source (source of the Chang Jiang, Huang He and Lancang Jiang) National Park, Northeast China Tiger and Leopard National Park, Giant Panda National Park, National Park of Hainan Tropical Rainforest, Wuyishan National Park
	Six nature parks involving terraces: Yunhe Terraces in Zhejiang Province, Nian Shan Terraces in Fujian Province, Yaopai Terraces in Guangdong Province, Longsheng Longji Terraces in Guangxi Zhuang Autonomous Region, Congjiang Jiabang Terraces in Guizhou Province, Honghe Hani Terraces in Yunnan Province
	Four national birds nature reserves: Dongzhai national nature reserve in Henan Province; Crested Ibis national nature reserve in Hanzhong, Shaanxi Province; Black-necked crane national nature reserve in Dashanbao, Yunnan Province; Black-necked crane national nature reserve in Yarlung Tsangpo River Midstream Valley, Xizang Autonomous Region
Key mineral resources	1. In the core reserve of protected areas, key mineral resources according to the regulations can be left
	2. Outside the core reserve of protected areas, some key mineral resources can be remained, others should be adjusted out the ERLs gradually
	3. For various natural parks, if legal mining rights are established in areas where exploration and mining are not forbidden by laws and regulations and do not cause obvious influence on ecological functions, the scope of protected areas can be adjusted accordingly and should not be included in the ERLs; those that have caused obvious influence should be withdrawn in order
	4. For newly classified protected areas and ERLs outside the protected areas, the mining rights that have been established according to the law should not be included in the ERLs
Artificial commercial forests	1. Considering the integrity of ecosystem and the stability of ecological patterns, the artificial commercial forests located in key ecological zones, such as areas along the banks of major river headwaters, around important wetlands and reservoirs, in areas severely affected by desertification and soil erosion, along the key coastal conservation forests and foundation forests, as well as scattered ones located in areas of critical ecological importance, should be included in the ERLs
	2. Other concentrated and contiguous commercial forests may not be included

(continued)

Table 2.1 (continued)

Conflicts type	Management rules
Major construction projects	Outside the core reserve of protected areas, the spaces occupied by major strategic and infrastructure projects such as transportation, energy and water conservancy, which have been clearly selected and have conformed to national development plan, the national special plan, the regional development strategy, in accordance with territorial and spatial planning at or above the provincial level and without significant influence on ecological functions, would not be delineated in the ERLs
Towns and villages	1. Urban areas are not in the ERLs 2. In the core reserve of protected areas, towns and villages are gradually moving out in order 3. Outside the core reserve of protected areas, administrative towns are not in the ERLs. The planning of villages is determined according to local conditions. Large and densely populated villages are not included in the ERLs and small scattered villages may be retained

geographical pattern and ecological functions, prioritizing the inclusion of extremely important ecological function areas, extremely fragile ecological areas and areas with potential important ecological value within the ERLs. On the other hand, the government strictly implements the priority of farmland conservation and seriously follows the ERLs to ensure national food security.

4. Seeking truth from facts. Insisting on problem-orientation, the Chinese government focuses on the coordination of development and security and properly handles the legacy issues, as well as minimizing new and unnecessary contradictions and conflicts to leave enough flexible space for future development. Respecting regional differences, the Chinese government delineates ERLs based on scientific evaluation, without pre-delineation indexes and decomposing tasks. The ERLs are matched with the features of physical geography pattern and spatial distribution of typical ecosystems in China. For example, in Xizang, Inner Mongolia and some other provinces (autonomous regions) the ERLs account for more than 50% of the territory area, but it is less than 10% of the total in Shanghai, Shandong, Jiangsu and Henan altogether.

2.2.2 Working Mechanism

1. Linking up the land use classification and sea area use classification with the unified database and base maps. The authority, accuracy and timely update of the basic data are the precondition for ensuring the strict implementation of the ERLs. The spatial data for the delineation uniformly adopts the 1985 national elevation datum and the geodetic coordinate system of 2000, using the results of the Third National Land Survey as the base data and base maps. The classification of territorial space is standardized, and the latest authoritative data of the

verification and rectification of permanent basic farmland and the shoreline survey are linked up, serving as the foundation for the delineation of the ERLs.

2. Determining the essential ecological function areas according to "Dual Evaluation". Through carrying out the "Dual Evaluation" scientifically, essential ecological function areas, such as water conservation, soil conservation, biodiversity conservation, windbreak and sand fixation and coastal conservation, as well as the most ecologically fragile areas such as soil erosion, rocky desertification, land desertification, coastal erosion and sand loss, are identified. Priorities are given to the above areas, so as to realize the integrated conservation of important ecological spaces.
3. Cooperation among departments at different levels and carried out multiple rounds of joint reviews. The MNR, MEE, NFGA and other departments jointly formed the working group. All departments jointly reviewed the delineation results of provinces (autonomous regions and municipalities directly under the central government) and put forward suggestions. They also carried out meetings to review the delineation status and conflict handling, etc. Led by the Chinese Academy of Sciences, an interdisciplinary expert group from natural resources, ecological environment, forestry and grass, water conservancy, transportation, residual building, rural area and other fields worked together on major issues in the process.
4. Ensuring the implementation by monitoring the quality of results. Work plans, requirements for achievement submission and key points of review were issued and released. Methods including polygon-by-polygon computer comparison, field survey, expert argumentation are involved. By communicating with the local government round-by-round, find problems and quickly rectify them in place. The plan of the ERLs in each province (autonomous region and municipality directly under the central government) should be confirmed by the people's government at the provincial level. The three control lines on the "one map" of territorial and spatial planning are well coordinated to ensure that there is no crossover, no overlap and no conflict. In accordance with the requirement on the consistency of "data, lines and map", the ERLs map with 1:10,000 scale accuracy are produced attached with the territorial planning map, recorded in the database, implemented on each parcel and firmly enforced by being applied in local spatial planning at all levels.

Chapter 3
Achievements

3.1 General Situation

The total area of China's ERLs reaches about 3.19 million km^2. Among them, the land ERLs cover an area of about 3.04 million km^2, accounting for more than 30% of the country's land area; the marine ERLs cover an area of about 150,000 km^2 (Fig. 3.1).

The land ERLs mainly consist of optimized and adjusted protected areas, extremely important ecological function areas, extremely fragile ecological areas and areas with potentially important ecological value. Among them, the protected areas account for about 56%, the essential ecological conservation areas outside the protected areas (extremely important ecological function areas, extremely fragile areas) account for 28% and areas with potentially important ecological values account for 16%.

The marine ERLs include the ERLs in the coastal management sea and the ERLs outside the coastal management sea, with a total area of about 150,000 km^2, 65% of which are nature reserves and others are key ecological function areas, such as coastal protection area and biodiversity conservation area, and extremely fragile ecological areas, like coastal erosion area and sand source loss area.

3.2 Spatial Distribution

3.2.1 Delineation Based on Territorial Regional Features

The delineation of the ERLs follows the principle of seeking truth from facts and fully respects the characteristics of physical geographical patterns and ecological functions in different regions. There are striking features regarding the quantity,

China Land Surveying and Planning Institute, *The Blue Book on China's Ecological Red Lines 2023*, https://doi.org/10.1007/978-981-95-0051-2_3

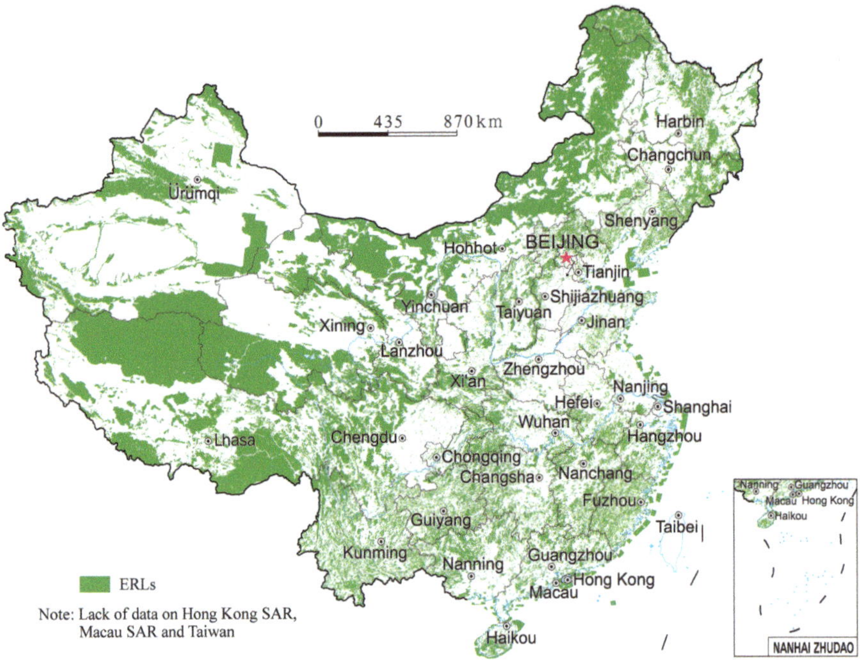

Fig. 3.1 China's ERLs

structure, spatial distribution and ecological function of the ERLs in different regions. In general, China's ERLs are mainly distributed in the western region, with an area of about 2.46 million km^2. The ERLs in the central region cover an area of 190,000 km^2, an area of 290,000 km^2 in the eastern region and an area of 250,000 km^2 in the northeast region. They account for 77%, 6%, 9% and 8% of the total area of the national ERLs (Fig. 3.2).

From the perspective of regional distribution, 80% of the land ERLs are distributed in eight provinces (autonomous regions)—Xizang, Inner Mongolia, Xinjiang, Qinghai, Heilongjiang, Sichuan, Gansu and Yunnan. The area of land ERLs in Shanghai, Tianjin, Beijing, Jiangsu and other provinces (municipalities directly under the central government) are less. Considering the proportion of the land ERLs in the total land area, Inner Mongolia, Xizang, Qinghai, Heilongjiang, Sichuan and other provinces (autonomous regions) have a relatively high proportion. Among them, Inner Mongolia and Xizang have more than 50% of their autonomous areas included in the ERLs. In Shanghai, Shandong, Jiangsu, Henan, Anhui and other provinces (municipalities directly under the central government), the delineation proportion is less than 15% (Fig. 3.3).

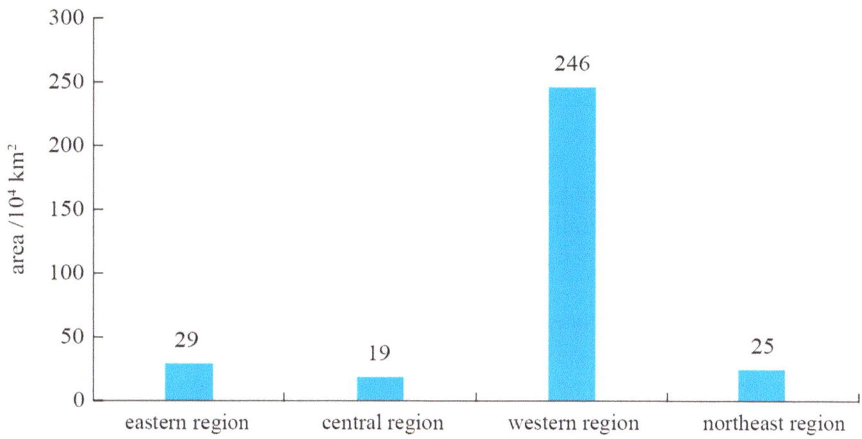

Fig. 3.2 Area of the ERLs in four regions of China (including land and marine)

Fig. 3.3 The proportion of China's ERLs in each province (autonomous region and municipality directly under the central government)

3.2.2 *Delineation Supporting Territorial Ecological Security*

The areas covered by ERLs are mainly distributed in the Qinghai-Xizang Plateau ecological zone, the Huang He key ecological zone, which includes the Loess Plateau ecological barrier, the Chang Jiang key ecological zone, including the Sichuan-Yunnan ecological barrier, the northeast forest belt, the northern sand prevention belt, the southern hilly mountain belt, and the coastal belt. The distribution has formed the "three zones and four belts" territorial ecological security barrier (Fig. 3.4).

The Qinghai-Xizang Plateau ecological zone is aimed at conserving the alpine ecosystem of the plateau, conserving the "China's Water Tower" and biodiversity, and improving climate change adjustment and water conservation of great rivers.

The Huang He key ecological zone is aimed at strengthening water conservation in the upper reaches of Huang He, restraining soil erosion in the middle reaches and strengthening the maintenance of ecological base flows and flood prevention in the downstream areas. It is also aimed at building ecological corridors connecting the Qinghai-Xizang Plateau, the Loess Plateau and the North China Plain to ensure high-quality development across the Huang He basin.

The Chang Jiang key ecological zone is aimed at strengthening water conservation and water and soil conservation in the upper reaches of the Chang Jiang, conserving the main stream and branches of the middle and lower reaches and lakes

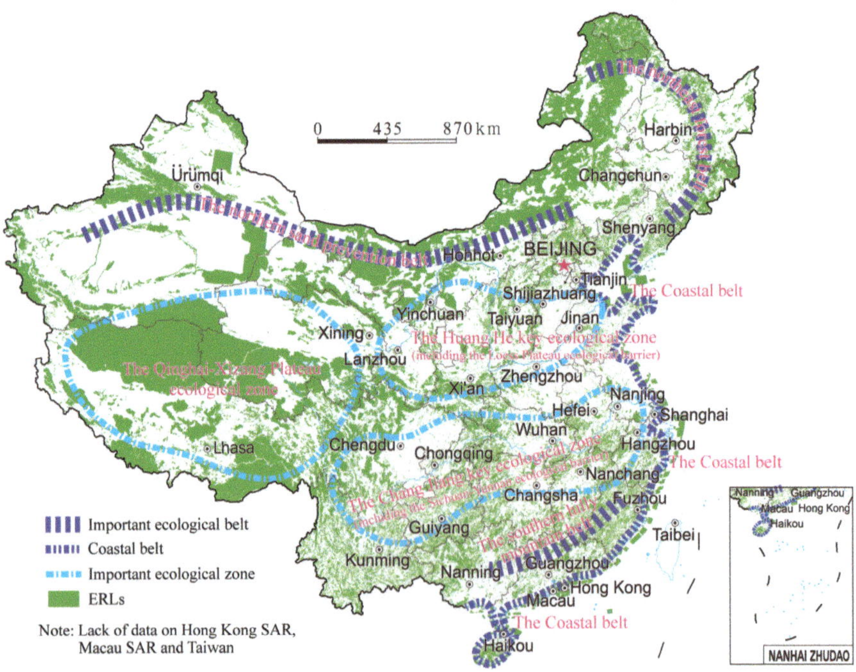

Fig. 3.4 Superposition of China's ERLs and the "three zones and four belts" ecological security barrier

and wetlands including Dongting Hu, Poyang Hu, Chao Hu, Tai Hu. It is also aimed at building the ecological corridors connecting the Qinghai-Xizang Plateau, Yunnan-Guizhou Plateau, the middle and lower reaches of the Chang Jiang plain to ensure the implementation of the ChangJiang Economic Belt Strategy.

The northeast forest belt is aimed at conserving the forest and wetland ecosystem in the northeast region and improving the water conservation capacity of the Songhua Jiang, Nen Jiang, Ussuri Jiang, Liao He, Yalu Jiang, etc.

The northern sand prevention belt is aimed at conserving the grassland and desert ecosystems in arid and semi-arid areas and improving the function of biodiversity conservation and windbreak and sand fixation, to ensure ecological security in the northwest, north and northeast regions.

The southern hilly mountain belt is aimed at conserving the sub-tropical forest ecosystem, strengthening the water conservation capacity of the Zhu Jiang and its southeast coastal rivers, comprehensively combating both soil erosion and rocky desertification.

The coastal belt is aimed at strengthening the ecological connectivity between the seaward and landward sides of the coastline and that between the ocean and river basins, and conducting the global conservation and systematic governance of watersheds-rivers-estuaries-coastlines-coastal wetlands-offshore areas-open sea ecosystems.

3.3 Structure of Land Use

3.3.1 First-Class Structure for Land Use

According to the results of the 2020 national land change survey, over 98% of the land area within China's ERLs is designated as ecological land use, including forest, grassland, wetland, water, sandy area, bare land (Fig. 3.5). Among them, forests, grasslands, wetlands, rivers, lakes and other areas with strong ecological

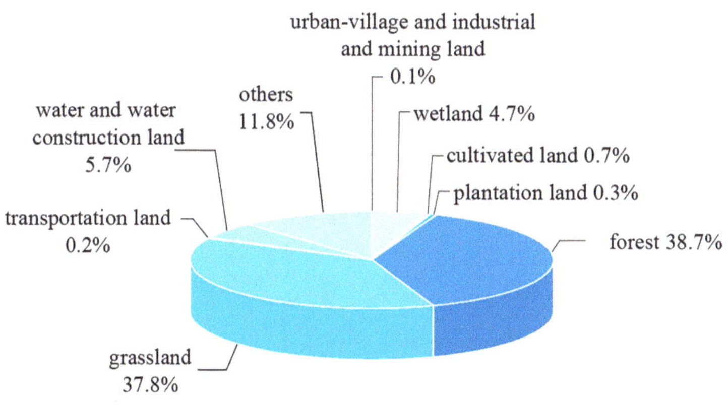

Fig. 3.5 First-class structure of land use within the ERLs

functions account for about 87%. In addition to ecological land use, the scope of the ERLs also includes a small amount of land for human activities, primarily for the livelihood of native residents, such as cultivated land, plantation land, transportation land, urban-village and industrial and mining land, and water construction land, accounting for about 1% of the total area of the ERLs.

3.3.2 *Second-Class Structure for Land Use*

From the second-class land use categorization, the main areas protected within the ecological land are those with the highest ecological quality and greatest value for conservation within the territorial space (Fig. 3.6). Within the ERLs, about 95% of the forest areas are arbor forestland and shrubland with high canopy density, and about 85% of the grassland areas are natural grassland, which can better fulfill ecological service functions, such as water source conservation, soil erosion control, carbon fixation, windbreak and sand fixation, and biodiversity conservation. Within the ERLs, marsh meadows, inland tidal flats, jungle marshes and marshes respectively account for 47.9%, 23.8%, 11.7% and 9.0% of the wetland area. In the water area, lakes account for 41.7%, glaciers and permanent snow for 26.9%, and rivers for 17.2%. It can enhance and maintain upstream water conservation functions, while breeding organisms and purifying the environment through conserving great rivers, lakes, and key water bodies.

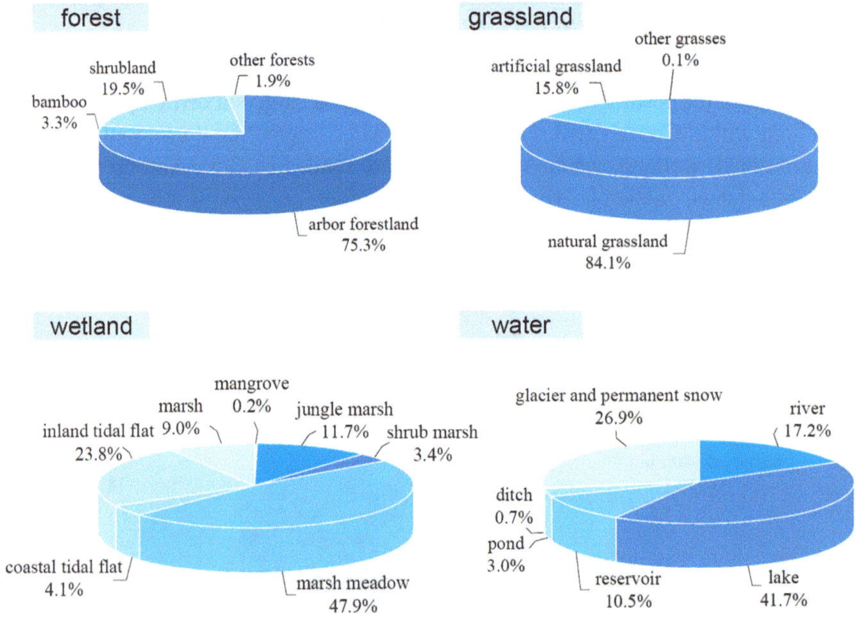

Fig. 3.6 Second-class structure of land use within the ERLs

3.4 Important Ecological Functions

The areas covered by the ERLs are those with special and important ecological functions within the scope of ecological space, which must be conserved compulsively and strictly. The delineation range is mainly located in essential ecological function areas with functions such as water conservation, biodiversity conservation, water and soil conservation, windbreak and sand fixation, coastal ecological stability, as well as the most ecologically fragile areas such as soil erosion, land desertification, rocky desertification, etc., and the areas with potentially important ecological value.

3.4.1 Giving Priority to Extremely Important Ecological Function Areas

In the delineation of the ERLs, based on the evaluation results of the importance of ecological conservation, the extremely important ecological function areas cover about 3.6 million km^2. After deducting artificial commercial forests, contracted pastureland, legal mining area, major projects area and other regions with human activities that do not meet the control requirements of the ERLs, all the extremely important contiguous areas of land ecological conservation are delineated into the ERLs, covering an area of about 1.73 million km^2. According to the evaluation results of the importance of marine ecological conservation, the extremely important area of marine ecological conservation in the coastal management sea covers about 72,000 km^2. After deducting the area with human activities that do not meet the control requirements of the ERLs, about 70,000 km^2 of coastal management sea is included in the marine ERLs.

3.4.2 Ensuring Ecosystem Service Functions

The extremely important ecological areas include areas with extremely important ecological function areas and extremely fragile areas. According to statistics, about 165.6 km^2 of extremely important ecological functional areas, with functions, such as water source conservation, biodiversity conservation, soil erosion control, windbreak and sand fixation, etc., are included in the land ERLs, accounting for about 54.5% of the total area of the land ERLs. The extremely important ecological function areas in the marine ERLs, which play important functions such as marine biodiversity conservation and coastal conservation, accounted for about 81.2% of the total area of the marine ERLs in the coastal management sea.

The extremely fragile ecological areas such as soil erosion, desertification, rocky desertification and sand loss within the land ERLs, cover about 219,000 km^2, accounting for 7.2% of the total area of the land ERLs. The extremely fragile areas such as coastal erosion and sand loss in the marine ERLs accounted for about 3.9% of the total area of the marine ERLs in the coastal management sea. Transferring these areas into the ERLs can minimize human interference in extremely important and fragile areas to ensure that the ecosystem plays important service functions.

3.5 Typical Ecological Classes

In the delineation results of ERLs, most typical ecological areas rich in biological diversity, such as the habitats for panda, Amur tiger, Chinese alligator, and other rare and endangered wild animals and plants, the source and upstream of important rivers, such as the Chang Jiang, Huang He, Lancang Jiang, Zhu Jiang, Min Jiang, Huai He and the Ergun He, as well as are important wetlands, glaciers and permanent snow, mangroves, seagrass meadows and coral reefs are primarily included in the ERLs. Among them, 99% of the mangroves, 96% of the first-class national non-commercial forests, 92% of the glaciers and permanent snow, 91% of the coral reefs, 89% of seagrass meadows and 94% of the undeveloped and uninhabited islands are included in the ERLs for conservation.

The ERLs cover 35 priority areas for biodiversity conservation in China and conserve more than 90% of typical ecosystems. It is beneficial to conserving the integrity of ecosystems, natural remains, natural landscapes and habitats of rare and endangered wild animals and plants, and to maintaining and restoring the quantity and habitats of rare and endangered wild animals and plants. It has an important role in conserving important ecological resources and improving ecosystem functions and service values.

3.5.1 Important Wetland

Wetland is the ecosystem with the highest primary productivity and the richest biodiversity on earth. It provides necessary habitat, migration, wintering and breeding sites for aquatic animals, aquatic plants, and a variety of rare and endangered wild animals, especially water birds. It plays important roles in conserving biological and genetic diversity.

In addition, wetlands have many ecological functions, such as regulating runoff and flood, supplementing groundwater, regulating the regional climate and fixing carbon dioxide, degrading pollution and purifying water quality. So far, there are 114 plots of wetlands have been included in the list of international and domestic important wetlands, mainly distributed in the Sanjiang Plain, the middle and lower reaches of the Huang He, the middle and lower reaches of the Chang Jiang, the Qinghai-Xizang Plateau, and coastal regions. All of them are included in the ERLs basically, and the control measures are implemented strictly.

Mangrove is a wetland woody plant community, which is naturally distributed in tropical and sub-tropical coastal intertidal zones, and consists of evergreen trees or shrubs with mangrove plants as the main body. It constitutes an important "blue carbon" resource in coastal waters together with seagrass meadows, coastal salt marsh wetlands and other ecosystems (Zhang et al. 2023b). Mangrove provides abundant food and good habitat for birds, fish and other marine organisms, and plays an important role in wave prevention and shore conservation, disaster prevention and reduction, air regulation and water purification.

According to the 2020 national land change survey, the total area of mangroves in China is about 271 km^2, mainly distributed in five provinces (autonomous regions), namely Guangdong, Guangxi, Hainan, Fujian and Zhejiang, 99% of which have been included in the delineation, totally about 268 km^2. Apart from scattered patches, which total area is about 3 km^2, all mangroves have been included in the ERLs area.

3.5.2 *National Non-commercial Forest*

National non-commercial forest refers to the shelterbelts and special purpose forests with an extremely important ecological location or an ecologically vulnerable condition. It plays an important role in land ecological security, biodiversity conservation and sustainable development of both the economy and society. It is aimed at playing the ecological and social service functions of forests.

According to the conservation level, the national non-commercial forest can be divided into first-class and second-class. China's first-class national non-commercial forests are about 170,000 km^2, mainly distributed in the northeast, Longnan region and key forest areas in the southwest region, such as the Great Khingan Ling, Lesser Khingan Ling, Qinba Shan, Hengduan Mountains. About 96% of them have been delineated in the ERLs this time. Apart from the patches that are relatively scattered and broken, located in the existing conflict areas such as urban and village construction land, legal mining areas, etc., the rest of the first-class national non-commercial forests with important conservation value are all delineated in the ERLs.

3.5.3 *Glaciers and Permanent Snow*

Glaciers and permanent snow refer to the land whose surface is covered by ice and snow throughout the year. As an important component of the cryosphere, glaciers and permanent snow are one of the important indicators of climate change. On the one hand, meltwater plays a role of "cutting peaks and filling valleys" through replenishing river flow. On the other hand, it is an important water resource for regional social and economic development (Zhang et al. 2023a). Therefore, the conservation of glaciers and permanent snow is related to the balance of regional water resources and the sustainable development of the social economy.

The total area of glaciers and permanent snow in China is about 50,000 km^2, mainly distributed in Xizang, Xinjiang, Qinghai, Gansu, Sichuan, Yunnan and other provinces (autonomous regions). About 92% of them have been delineated in the ERLs this time.

3.5.4 Coral Reef

Coral reef is formed by the continuous accumulation of the calcified skeletal remains of coral polyps, calcified algae, foraminifera and other reef-building organisms. It is one of the most biologically diverse and productive ecosystems in the marine environment, rightfully called an "oasis in the ocean". The coral reef ecosystem can provide mankind with rich fishing and tourism resources and conserve the coastline (Li et al. 2022).

In China, coral reefs are mainly found in coastal areas of Hainan, Guangdong, Guangxi, Fujian and other provinces (autonomous regions and municipalities directly under the central government). This time, about 91% of coral reef areas are delineated in the ERLs, which is aimed at improving the conservation and restoration of the coral reef ecosystem through reducing the disturbance of human activities.

3.5.5 Seagrass Meadow

Seagrass usually grows in waters at a depth of 6 m below sea level and forms seagrass meadows in large patches, which is one of the most important coastal ecosystems in the world. It can provide food and habitat for many marine organisms (Xu et al. 2022).

Seagrass meadow has the function of carbon reduction, oxygen production, and climate regulation. It can also effectively absorb and fix carbon, nitrogen, phosphorus and other substances in seawater, and purify the water environment, serving as an important marine carbon sink. At the same time, the seagrass meadow functions as a habitat, feeding ground and nursery ground for marine organisms with high productivity, but it can also weaken waves and ocean currents (Han et al. 2023). Affected by human destruction, environmental pollution, natural disasters and other factors, however, it is estimated that the rate of seagrass meadows degradation globally is five times that of tropical rainforests.

At present, seagrass meadows in China's coastal management area covers about 106 km^2, 89% of which are included in the ERLs, mainly distributed in the Yellow Sea and Bo Hai areas, which are dominated by the Caofeidian sea area in Hebei Province, Changhai County sea area in Liaoning Province, and the southern sea areas such as coastal areas of Guangdong, Guangxi and Hainan provinces (autonomous regions).

3.5.6 Uninhabited Island

Uninhabited island refers to islands that are not registered as residential addresses under the household registration management system. As a unique marine resource, uninhabited islands not only fulfill important ecological functions such as coastal conservation and water source conservation, but also are of great importance in safeguarding the ecological security and maritime rights and interests of the islands, due to their special geographic positions (Li et al. 2023).

About 6,800 uninhabited islands have been delineated in the marine ERLs, including special planning islands, uninhabited islands with rare and endangered wild plants and their habitats, important natural relics and other special conservation values, as well as ecologically important uninhabited islands identified through the evaluation of resource and environmental carrying capacity. The number of the uninhabited islands included in the ERLs accounts for 64% of the total in China, and 94% of those that have not been developed in China.

Chapter 4
Management and Control Measures

4.1 Clarifying Control Rules

4.1.1 Central Government Level

The ERLs encompass a diverse array of protected areas, including national parks, protected areas and nature parks, as well as important ecological spaces such as important wetlands, first-class national non-commercial forests, first-class conservation areas of drinking water sources, important estuaries, mudflats and shallow sea waters (Table 4.1). China's legislative system addresses these areas through a suite of pertinent laws and regulations such as *Wetland Conservation Law of the People's Republic of China, Forest Law of the People's Republic of China, Marine Environmental Conservation Law of the People's Republic of China, Regulations of the People's Republic of China on Protected Areas, Tourism Law of the People's Republic of Chin*a, establishing a solid foundation for an all-inclusive legal framework that ensures rigorous management and control of the ERLs.

Based on the current laws and regulations, the delineation of the ERLs followed the principle of the combination of zoning and controlling. In 2019, the Chinese government issued documents, specifying the principles and requirements of the construction of the protected areas system and the delineation of three controlled lines. In 2020, MNR and NFGA jointly issued guidelines that clarified the rules for optimizing adjustment and handling contradictions and conflicts in protected areas. In 2022, MNR, together with MEE, NFGA issued a policy that standardized the specific types and management requirements of limited human activities within the ERLs, and the specific circumstances of national major project occupation and approval procedures, and clarified the regulatory responsibilities of relevant departments (Table 4.2).

China Land Surveying and Planning Institute, *The Blue Book on China's Ecological Red Lines 2023*, https://doi.org/10.1007/978-981-95-0051-2_4

Table 4.1 The constitution of the ERLs in China

<table>
<tr><th colspan="3">The constitution of the ERLs</th><th>Main ecosystems and ecological regions</th></tr>
<tr><td rowspan="5">Land</td><td rowspan="3">Land protected areas</td><td>National park</td><td rowspan="5">First-class national non-commercial forests, important wetlands, glaciers and permanent snow, natural shorelines, distribution habitats of species with extremely small populations, first-class conservation areas of drinking water sources, etc.</td></tr>
<tr><td>Nature reserve</td></tr>
<tr><td>Natural park</td></tr>
<tr><td colspan="2">Extremely important ecological function areas, extremely fragile ecological areas</td></tr>
<tr><td colspan="2">Areas with potentially important ecological value</td></tr>
<tr><td rowspan="4">Marine</td><td rowspan="2">Marine protected areas</td><td>Marine nature reserve</td><td rowspan="4">Important estuaries, specially protected islands, distribution areas of rare and endangered species, mangroves, coral reefs, seagrass meadows, growth and breeding areas of fishery resources, important coastal salt marshes, important mudflats and shallow sea waters, important seaweed fields, uninhabited islands, etc.</td></tr>
<tr><td>Marine park</td></tr>
<tr><td colspan="2">Extremely important ecological function areas, extremely fragile ecological areas</td></tr>
<tr><td colspan="2">Areas with potentially important ecological value</td></tr>
</table>

4.1.2 Local Government Level

According to clear management requirements of the ERLs at the central government level, combined with the actual situations, local governments should insist on the combination of principle and flexibility, and refine the control requirements of the ERLs. So far, the control rules of the ERLs have been issued in Zhejiang, Jiangxi, Shanghai, Shandong, Anhui, Sichuan, Guizhou, Shaanxi, Hainan, Guangxi, Tianjin and other provinces (autonomous regions and municipalities directly under the central government).

4.2 Improving Management Measures

4.2.1 Administration According to Different Situations

According to the principle of no crossing, overlap or conflict of the three controlled lines, it is recommended to explore measures to adopt classified management for special situations.

For example, some rare birds depend on crops, in order to provide the conservation of bird habitat, the surrounding permanent basic farmland can be included in the scope of protected area, as well as combined with the dynamic optimization of the boundary of the future nature reserve, meanwhile, the general farmland in the protected area can be gradually included into the permanent basic farmland protect area combined with the adjustment and rezoning of permanent basic farmland to

Table 4.2 Management required for the ERLs

Management principles	Management requirements
Strict standardizing the limited human activities	1. In principle, human activities are prohibited in the core reserves of protected areas within the ERLs 2. Outside the core reserves of protected areas, developmental and productive construction activities are strictly forbidden. Under the premise of following laws and regulations, only limited human activities that do not disrupt ecological functions are allowed (not regarded as occupying the ERLs). For the application for the new land, sea and island use, it is necessary to submit the opinion of the people's government at the provincial level about the approval for the limited human activities within the ERLs. For those which don't involve the examination and approval of new construction land and islands and sea for use, it is necessary to manage according to the related regulations. If there are no clear regulations, the people's government at the provincial level should formulate specific supervision methods
Strict approval for the occupancy of the ERLs	In addition to the limited human activities permitted, the national major projects that need to occupy the ERLs (excluding the newly increased land reclamation and island use) should be pre-reviewed by the MNR for land or marine use according to regulations and then file to the State Council for approval. For the application of land or marine use for approval, it must be accompanied by a demonstration opinion from the provincial people's government, which justifies the necessity of occupying the ERLs and describes the measures of intensive land use and reducing the impacts of ecological environment, based on the "one map" of territorial and spatial planning and the use control requirements For the limited human activities permitted within the ERLs and major national projects which occupy land temporarily, it is necessary to strictly implement the restoration responsibility according to the relevant requirements of MNR on standardizing the temporary land use, with reference to the regulations of temporary land occupation of permanent basic farmland
Handling the issues left over from history in good order	1. After the ERLs are approved by the State Council, the provincial government should formulate withdrawal plans according to the principle of respecting history and seeking truth from facts, and clarify the scheduling, compensation and resettlement, ecological restoration and other requirements, thus ensuring ecological security and social stability 2. If conditions permit, the artificial commercial forests should be managed by means of leasing, replacement and redemption. The artificial commercial forests in important ecological areas should be gradually converted into non-commercial forests according to provisions 3. Hydropower, wind power, photovoltaic and marine energy facilities that are distributed sporadically should be managed according to relevant laws and regulations. It is strictly prohibited to expand its area. At the expiration of the project, the construction unit is responsible for the ecological restoration

explore the "dual" management of the ERLs and permanent basic farmland in the future.

For another example, for the terraced wetland parks such as Hong He Hani Terraces in Yunnan Province with compound functions of ecological function as well as farming and cultural function, exploring the dual management of the ERLs and permanent basic farmland in the future can be considered on the basis of full respect for local intention.

4.2.2 Exploring Adaptive Management Measures

To reduce contradictions and conflicts in the process of delineation of the ERLs, permanent basic farmlands, mining rights, major construction projects, and artificial commercial forests are transferred out of the ERLs. In the next step, it is recommended to explore the implementation of ecological control measures in these areas in order to reduce the disturbance of human activities to ecosystems.

For example, for the permanent basic farmland around the ERLs, it is necessary to guide farmers to implement ecological planting and strictly control the use of pesticides and fertilizers; for exploration and exploitation of mineral resources, it is necessary to strictly evaluate the impact of mineral development on ecosystems according to the requirements of green mine construction, and to strengthen daily supervision, regular inspections, and timely ecological restoration; for artificial commercial forests, it is necessary to adopt more scientific and rational logging methods and renewal technologies to minimize the influence on ecosystem functions.

4.2.3 Regular Monitoring and Evaluation

An accurate and efficient evaluation system for regular monitoring and effectiveness evaluation of the ERLs can provide important support for strict adhering to the ERLs. It is recommended that MNR, together with relevant ministries and commissions, study and carry out technical standards for ERLs monitoring and performance, then conduct the national ERLs monitoring and performance effectiveness evaluation. Based on the investigation and monitoring system of natural resources, combined with the requirements for the implementation and evaluation of territorial and spatial planning, the monitoring index system of the ERLs should be established from the aspects of the study of the structure, pattern, quality, function and risk of the ecosystem.

It is necessary to monitor the changes in land use and human activities within the scope of the ERLs, and analyze the change trend in the ecological status of the ERLs. It is necessary to explore and carry out the performance effectiveness evaluation including the ecological status evaluation and ecological risk early warning of the ERLs. Besides, it is vital to implement evaluation, optimize the spatial layout of the ERLs properly, and ecosystem conservation and restoration through combining with the resource and environmental carrying capacity monitoring and early warning, as well as the territorial and spatial planning.

4.3 Establishing and Improving Long-Term Mechanisms

A sound long-term control mechanism is the fundamental guarantee for the ERLs' maintenance and sustainability. For the long-term control of the ERLs, balance must be considered across different regions, among interest groups, and between ecological security and regional socio-economic development. This is achieved by strengthening administrative regulations, constructing a technology system for control, and ensuring that protective measures are in place (Fig. 4.1).

4.3.1 Speeding Up the Legislation

Sound rules and regulations are the basics for strictly following the ERLs. Relevant laws of the ERLs need to be improved. The ERLs, due to its extensive coverage and the complexity of interests among various parties, including various departments at central and local government levels, individuals, other subjects, necessitates more explicit and comprehensive legal guarantees in the aspects of conservation responsibility, control measures, accountability system, and the investigation and punishment of illegal activities.

In the amendments of *National Park Law, Regulations of the People's Republic of China on Protected Areas, Tourism Law of the People's Republic of China,* etc., it is necessary to ensure that the regulations of the protected areas are consistent

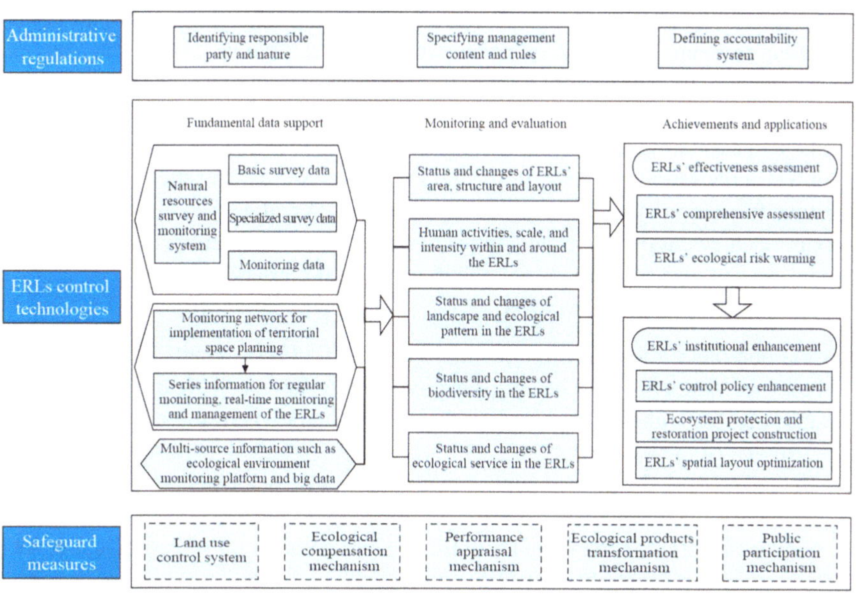

Fig. 4.1 Institutional framework of the ERLs

with the ERLs control rules, rationalize the relationship between the ERLs and the laws and regulations of different regions. At the same time, laws and regulations on ERLs management should be issued as soon as possible. It is necessary to insist on both principle and flexibility, fully considering the features of different types of regions, different human activities and their influence on the ecological environment, determine the ERLs control contents, rules and methods scientifically and rationally, so as to ensure a legal basis for ERLs supervision.

4.3.2 Improving the Compensation Mechanism

Ecosystems benefit from the maintenance and adjustment of climate, air and water resources widely, with the feature of regional spillover effects; water conservation, biodiversity conservation, soil erosion control, windbreak and sand fixation and other ecological functions also show the special properties of no-boundary flow. This means that there is not always a corresponding relationship between ecological protectors and beneficiaries.

China's ERLs are mainly distributed in the western provinces (autonomous regions) of Inner Mongolia, Xizang, Ningxia, Qinghai, Gansu and Guizhou. The governments of these regions not only assume the strategic responsibility of safeguarding the ecological security border and building a strong ecological security barrier, but also face the urgent need for developing the economy and improving people's livelihood. At the same time, these provinces (autonomous regions) are the concentrated distribution areas of ecologically vulnerable areas in China, which are characterized by poor stability of the ecosystem, relatively weak self-repair ability, and great pressure on ecological conservation and restoration.

To balance the relationship between the subject of ecological responsibility and the subject of beneficiary, and balance the relationship between ecological conservation and economic development, it is necessary to improve the ecological conservation compensation mechanism for the ERLs control, expand the path to achieve the transformation of natural resources to wealth and realize the value of ecological products, carry out the rights and duties of ecological conservation, and improve the enthusiasm of all parties to participate in ecological conservation.

The suggestions are as follows:

On the one hand, the government should regard the ERLs results as an important basis for national financial transfer payments to key ecological function areas, and increase transfer payments to the stakeholders where there is large—scale coverage of the ERLs.

On the other hand, the government should explore the establishment of a horizontal compensation mechanism between the areas where the ERLs are located and the beneficiary areas, enhancing the initiative and enthusiasm of ecological conservation areas through interest adjustment.

4.3.3 Establishing a Dynamic Adaptation Mechanism

As an innovative system, the results of the ERLs are based on comprehensively considering the current physical geographical features, the importance of ecological functions, ecosystem services and the needs of social and economic development.

After the ERLs delineation, according to the overall territorial and spatial planning at all local levels, the government should formulate the implementation of the ERLs, and further optimize the spatial distribution of the ERLs and surrounding areas. For scattered villages within the ERLs, it is necessary to compile according to the urban and village territorial and spatial planning, scientifically assess the current status of villages, reasonably divide village categories, and implement classified guidance.

Uphold the ERLs' integrity and stability through regular physical examination and evaluation integrated with territorial and spatial planning, so as to improve the carrying capacity and ecological functions of the ecosystem. In the adjustment of territorial and spatial planning, the newly added ecological restoration projects of mountains, rivers, forests, farmlands, lakes, grasslands, sands, and the essential ecological function areas that meet the control rules after the withdrawal of human activities are added into the ERLs to improve the integrity and connectivity of the ERLs. At the same time, it is necessary to adjust and establish a linkage working mechanism with protected areas to ensure the consistency of protected areas and ERLs, and update the "one map" of territorial and spatial planning in real time.

4.3.4 Improving the Public Participation Mechanism

A good ecological environment is the fairest public product and the most inclusive welfare for the people. The ERLs are not only related to safeguarding national and regional ecological security, but also related to the quality of people's livelihood and social harmony and stability. The effects of the ERLs implementation depend on the active and broad participation of the public.

To establish and improve the public participation mechanism, it is necessary to give full play to public service advertising, nature education, promote all sectors of society to firmly establish the concept of the ERLs, disclose information on the distribution, monitoring and conservation status of the ERLs to the whole society in time, thus leading to the whole society's fully understanding of the importance of delineation and strictly following the ERLs. Also, stimulating the public's self-awareness and internal vitality for ecological conservation and improving people's sense of acquisition and identity with ecology are important, so as to form a good situation of co-conservation, co-governance and sharing.

Chapter 5
Practical Cases of Local Governments

5.1 Jiangxi Province: Overall Balance Between Development and Ecological Conservation[1]

As the first batch of National Ecological Civilization Pilot Areas, Jiangxi Province has implemented the strictest ecological environmental conservation system based on the construction of a National Ecological Civilization Pilot Area. Meeting the national deployment requirements, the provincial government carried out overall conservation and development, scientific and reasonable delineation, actively explored the construction of management system and strived to realize the goals of making its layout more reasonable, delineation more accurate and management more effective.

5.1.1 Practice in Delineation

5.1.1.1 General Thought

Jiangxi Provincial Government gave priority to ecology and conservation by unifying the base maps on the basis of the data from the Third National Land Survey. The government carried out the "Dual Evaluation", identified extremely important

[1] This part was written by Wu Xin, Wang Jianhui, Li Tonghua, Guo Song and Yu Shaokang. Wu Xin, the director of the Territorial Space Planning Bureau, the Department of Natural Resources of Jiangxi Province; Wang Jianhui, the deputy dean of Jiangxi Provincial Territorial Spatial Survey and Planning Institute; Li Tonghua, an engineer of the Geographic Information Engineering Team of Jiangxi Provincial Geological Bureau; Guo Song, an engineer of Jiangxi Provincial Institute of Territorial Space Survey and Planning; Yu Shaokang, a professor-level senior engineer of Jiangxi Provincial Institute of Territorial Space Survey and Planning.

China Land Surveying and Planning Institute, *The Blue Book on China's Ecological Red Lines 2023*, https://doi.org/10.1007/978-981-95-0051-2_5

ecological function areas and extremely fragile ecological areas, and took the important ecological spaces such as protected areas, first-class conservation areas of drinking water sources and important wetlands into the scope of the ERLs to strengthen the bottom line of ecological security.

On the premise of following national regulations, combined with the actual situation, Jiangxi Provincial Government sorted out and formulated the adjustment rules that suit its features, properly handled various contradictions & conflicts within the ERLs, coordinated development and security, realistically reserved space for economic and social development, scientifically and rationally delimited the scope of the ERLs.

5.1.1.2 Delineation Principles

During the delineation process, Jiangxi Provincial Government insisted on the following principles:

First, upholding the bottom line and prioritizing ecology. Jiangxi Provincial Government insisted on the guidance of safeguarding territorial ecological security, systematically analyzed and determined the condition of ecological conservation, ensured that important ecological spaces were fully conserved and delineated to realize the goal of "Guarding the boundary of natural ecological security by the ERLs".

Second, insisting on scientific objectivity and seeking truth from facts. Based on reality, Jiangxi Provincial Government properly handled the contradictions & conflicts between boundaries of the ERLs and farmland, permanent basic farmland, mining rights, artificial commercial forests, towns and villages, and major projects in strict accordance with the principle of respecting history and seeking truth from facts. Through carrying out scientific and reasonable evaluations, the ERLs were adjusted scientifically. The location and the proportion of the ERLs were not determined from the beginning.

Third, insisting on overall planning, coordination and balanced development. On the premise of ensuring ecological safety, ecological function and the integrity of the ecosystem, the provincial government coordinated the relationships of ecological conservation, economic and social development, and properly reserved space for economic and social development.

5.1.1.3 Technology Flowchart

The technology flowchart is shown in Fig. 5.1.

1. Choosing pilot cities to test the rules at first. According to the idea of "exploring experiences by pilot cities, then promoting to the whole province", Jiangxi Provincial Government selected 11 counties (cities) with typical problems, including Jing'an County, Xunwu County and Chongyi County as pilots; sys-

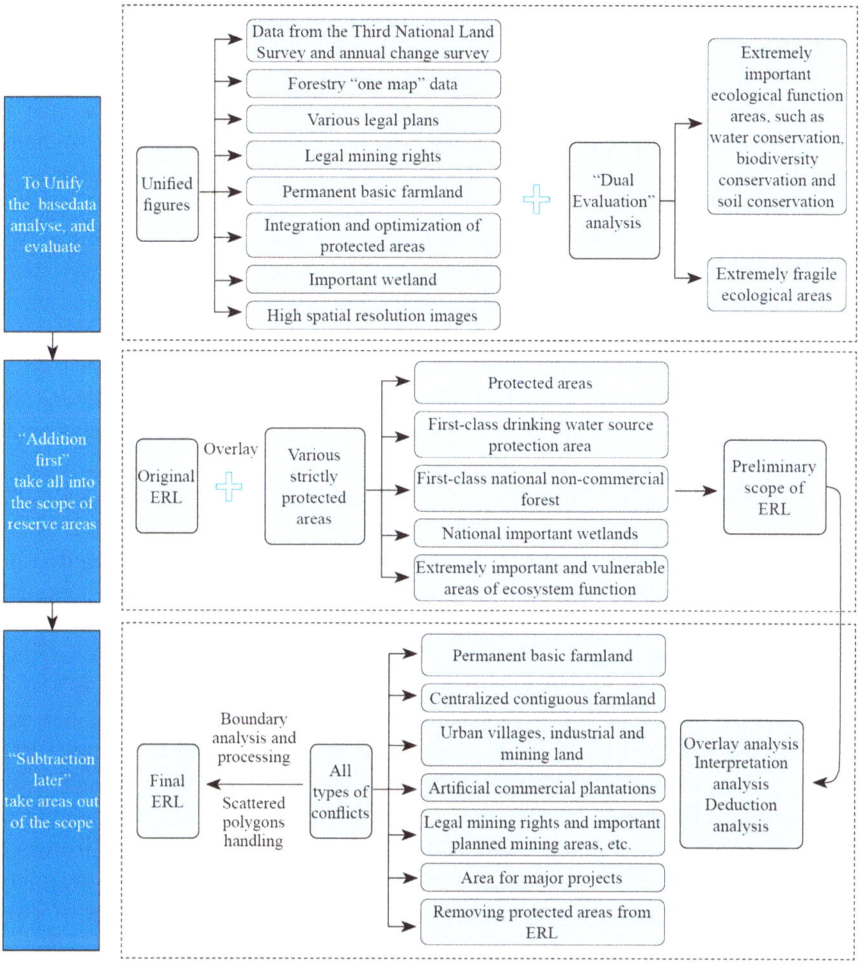

Fig. 5.1 Technology flowchart for evaluation and adjustment

tematically analyzed the causes of contradictions & conflicts by the means of "interpretation indoor at first and then field investigation". The interviewers met with the relevant local departments and farmers through symposium to find out conflicts and problems, then summarized a set of technical rules for how to do the ERLs evaluation and adjustment rules of Jiangxi Province.

2. Using the same set of maps and figures so as to consolidate the "base" work. The consistency of the maps and the figures was the foundation for scientifically delineating the ERLs. To coordinate and solve the problems such as the large number of data parameters of the base map and the inconsistency of the projection coordinate system, Jiangxi Provincial Government sorted out and checked the basic data such as the Third National Land Survey and the Annual Change Survey, the "Dual Evaluation", the "One map" of forestry, the map of the first-

class national non-commercial forest, the map of important wetlands, the annual high-definition remote-sensing images, planning data, mining rights and other required data for the evaluation and adjustment of the ERLs, then verified the legal compliance of data sources, converted coordinate parameters, in order to ensure the reliability of basic data, unified the base map and base data, strengthened data support and consolidated the "base" for scientifically and accurately delineating the ERLs.

3. Doing addition at first. Taking all important ecological spaces into the scope of the ERLs, so as to consolidate the ecological base. Jiangxi Provincial Government has systematically sorted out the extremely important ecological function areas, extremely fragile ecological areas, protected areas and other important ecological spaces, gave priority to conservation to ensure those areas are protected as much as possible. The extremely important ecological function areas and the extremely fragile ecological areas should be delineated into the ERLs as much as possible.

   Jiangxi Provincial Government has carried out "Dual Evaluation" to identify the essential ecological function areas with important water conservation, biodiversity conservation and soil conservation, and also the extremely fragile ecological areas of soil erosion control. The contiguous extremely important ecological conservation areas without conflicts were included in the ERLs. The scope of integrated and optimized protected areas has been fully included in the ERLs.

   Jiangxi Provincial Government simultaneously carried out the evaluation and adjustment of the ERLs, which classified all the protected areas into the ERLs scope in accordance with the integrated optimization plan of protected areas. After the adjustment, the protected areas within the ERLs have increased from 12,810.64 to 15,899.21 km^2, and the proportion of included areas has increased from 77.92% to 100%, which has further increased the conservation strength of protected areas and maintained the ecological security pattern of the province.

   Other areas that require strict conservation have been included in the ERLs as much as possible. Combined with relevant data of various departments, Jiangxi Provincial Government has included the first-class national non-commercial forests, first-class protected areas for drinking water sources, important wetlands and other areas with important ecological service functions, potentially important ecological values, and areas that need to implement strict conservation into the ERLs, so as to maintain the integrity and systematicity of ecosystem service functions.

4. Doing subtraction when facing contradictions & conflicts, and taking overall consideration for development space. According to the function structure of the ERLs, Jiangxi Provincial Government handled the contradictions within the ERLs in different areas.

   After the process of integration and disposal of the conflicts, the protected areas were fully included in the ERLs. The contradictions in the ERLs but outside the protected areas were handled according to the following methods:

   Respecting geographical patterns, coordinating and ensuring food security. Considering the mountainous and hilly landform and the features of "small and scattered" distribution of farmland, Jiangxi Provincial Government classified the farmland by polygon distribution and total areas. In principle, the contiguous farmland of over 5 mu[2] and the permanent basic farmland of over 3 mu were adjusted out of or not included in the ERLs while also considering farmland conservation.

Respecting development reality, coordinating and reserving space for development. Considering the development needs of towns and villages and the possibility of reasonable expansion, as well as the avoidance of risks that may be difficult to manage if the delimitation is performed according to map polygons only, the polygons of towns and villages and industry & mining construction land of more than 1 hectare in a row and the outer buffer zone of 50 m were adjusted out of or not classified into the ERLs, so as to reserve some buffer spaces for development and management. At the same time, Jiangxi Provincial Government insisted on overall development and security. Major projects (engineering) at or above the provincial level that had been conformed by the territorial and spatial planning at or above the provincial level, and the location has been sited, the plots were adjusted out of or not classified into the ERLs, so as to reserve the space for the construction of major projects and reasonably ensure high-quality social and economic development.

Respecting the legal rights and interests and coordinating and ensuring local sources of income. Mining and forestry are two main social sectors of Jiangxi Province and remain important pillars of the province's economic sources for the local villagers. To protect the legal rights and interests of relevant stakeholders, the licensed mining rights, exploration rights and the approved important planning mining areas, orefields, as well as artificial commercial forests outside the source areas of five rivers, were either adjusted out of the ERLs or not classified into them to ensure the legal economic income of local governments.

5.1.2 Exploration and Practice of Administration System

Jiangxi Provincial Government has actively promoted the "combination of delineation and management" of the ERLs and successively issued policies on overall delineation and implementation of three controlled lines in territorial and spatial planning and other documents to strengthen the approval of construction activities within the ERLs and preliminarily established the ERL management system.

1. Strengthening the bottom line constraint and seriously safeguarding the seriousness and authority of the ERLs.

   First, arbitrary adjustment of the scope of ERL is forbidden. The ERLs may not be arbitrarily adjusted once the results of the ERL delineation are approved. Local governments regarded the ERLs as an important basis and precondition

[2] 1 mu ≈ 666.67 m^2.

for comprehensive decision-making. Various construction projects give priority to avoiding the ERLs; if it is impossible to avoid, it must meet the requirements for the management of the ERLs so as to resolutely uphold the seriousness and authority of ERLs.

Second, the ERLs were included in the approval matters of construction land application. In the process of approving the construction, Jiangxi Provincial Government regarded the ERLs as the mandatory review content, strictly reviewed the use, layout, compliance, necessity and other contents of construction projects and implemented the "one-vote-veto" system for construction projects that involve the ERLs and do not meet the control requirements.

Third, a joint examination mechanism for the approval of construction projects within the ERL was established. In the process of land pre-examination and site selection, land application for approval, the nature resource authority of the province asked for opinions from the ecological environment and forestry departments at provincial level, had them participate in the relevant demonstration review, acquired clear opinions on whether the construction project meets the requirements of protected areas, forest land occupation, drinking water source, ecological environment conservation, etc. For a construction project which does not conform to the relevant laws and regulations, land application for approval shall not be accepted.

2. Setting strict working requirements and making solid progress in identifying human activities within the ERLs.

In November 2022, Jiangxi Provincial Government issued the Guideline on strengthening the management of the ERLs, and specified the specific requirements for manual activities approval and strengthened the supervision and management of the ERLs.

First, strengthening the necessity for site selection. In the process of selecting sites for construction projects, all localities and departments gave priority to avoiding the ERLs. For those that meet the control requirements and cannot avoid the ERLs, construction project units or industry authorities should strengthen the necessity of project construction, the rationality of site selection, and the comparison of several options for site selection, and explain the inescapability. Only the project that meets requirements can go through the approval procedures for admission.

Second, strengthening classified approval. When construction land demand arises, the county government should apply to the provincial government to issue a determination opinion on allowing limited human activities within ERLs. Among them, for the national major projects that really need to occupy the ERLs, the provincial department of natural resources should organize the demonstration, and apply to the provincial government for the issuance of the determination opinion on the inescapability of the ERLs and report to the State Council for approval.

Third, strengthening the responsibilities of each party. When handling the determination opinion, the municipal government should issue the preliminary determination opinion on allowing limited human activities within ERLs. In the construction project land pre-examination and site selection, the application for agricultural land conversion and land expropriation is submitted along with the determination opinion. If national parks, protected areas, tourist attractions and other protected areas, important wetlands, drinking water source reserves and other areas are involved, the relevant competent departments should give a definite opinion step by step.

3. Adopting information technology to improve the management efficiency of the ERLs.

   First, Jiangxi Provincial Government has incorporated the scope of the ERLs and relevant information into the "one map" implementation supervision information system for territorial and spatial planning and connected with the Approval System of Construction Land use, Review System of Development Plan and the Certificated System of Mining Rights, so that all departments can rapidly check the compatibility with the ERLs.

   Second, Jiangxi Provincial Government has distributed the implementation supervision information system account of territorial and spatial planning "one map" to all relevant departments, so that relevant departments can invoke the scope and information of ERL and provide services in planning, layout, project location and other decisions, and give priority to avoiding the ERLs.

   Third, Jiangxi Provincial Government has incorporated the scope of the ERLs and relevant information into the "Jiangxi Survey Cloud" system and opened the system to the public. The public can check whether planned urban development, mining, construction project layout, planning and layout and the route selection of road are involved in ERLs anytime and anywhere in order to avoid the ERLs, basically and effectively.

5.1.3 Strengthening Management in the Future

The "first half of the article" on the ERL delineation has been successfully completed and the ERLs are more accurate and the layout is more reasonable. How to systematically do a good job of the remaining work, that is, how to achieve the scientific management and protection of the ERLs, is still a major issue that should be solved in the future.

1. Promoting the "legalization" of supervision. At present, the laws and regulations issued by the state have no clear requirements for the supervision and implementation of the ERLs. Moreover, the existing laws and regulations in various areas within the ERLs cannot be integrated. As a result, the legal status remains to be clarified.

   Jiangxi Provincial Government should issue laws and regulations on the management of the ERLs as soon as possible, clarify the supervision responsibilities of the departments and straighten out the relationship between the management of ERLs and relevant internal regional laws and regulations, and define the law enforcement regulations on the violation of the ERLs, so as to realize the management and the supervision of the ERLs in accordance with a unified law.

2. Promoting the disclosure of the ERLs information. At present, the boundary of the ERLs has not yet been fully marked on the ground. Unauthorized occupation may occur.

   Jiangxi Provincial Government should start the boundary marking on the ground as soon as possible, adopt the combination of electronic fences and physical boundary piles, set up boundary stakes and signs in areas with frequent human activities, strengthen warnings through electronic fences, and establish a convenient working platform for the ERLs scope visualization, query, thus enabling the public to more conveniently inquire about and avoid the ERLs and enhancing the real-time conservation of the ERLs.

3. Promoting the "scientific modernization" of the ERLs supervision. At present, the related works of the ERLs focus on the delineation, but the evaluation and monitoring are still insufficient.

   Jiangxi Provincial Government should improve the technical system for ERL evaluation and monitoring, formulate long-term evaluation and monitoring indicator systems and technical methods for ecological environment quality and human activities within the ERLs, and establish a "monitoring-evaluation-early warning" mechanism to help management departments in scientific decisions.

   At the same time, Jiangxi Provincial Government accelerated information sharing and exchanging, constructed a joint supervision channel which combined by the One Map: Implementation Supervision Information System for Territorial and Spatial Planning and the Ecological Environment Supervision Platform, thus forming a regulatory synergy.

5.2 Hainan Province: Comprehensive Management of Land and Marine[3]

Hainan Province is the only tropical island province in China, which is characterized by superior climatic conditions and the highest biological species and endemic populations in China.

Hainan Province is the natural treasure vault and resource base of China's biodiversity and also the natural gene pool of China and even the world. In addition to the unique forests, rivers, grasslands, scrubs and other land biodiversity resources, Hainan Province also has coastal wetlands, mangroves, coral reefs, seagrass meadows, lagoons and other marine ecosystems, with the extremely rich biodiversity of islands and oceans.

[3] This part was written by Wang Jungang, Li Jia and Chen Tianping. Wang Jungang, the chief planner of the Department of Natural Resources and Planning of Hainan Province; Li Jia, the director of the Global Planning Office, the Department of Natural Resources and Planning of Hainan Province; Chen Tianping, a grade IV researcher of the Global Planning Office, the Department of Natural Resources and Planning of Hainan Province.

The mountainous areas in the central and southern parts of Hainan Province are important ecological function areas throughout the country, as well as source areas of major rivers, important water conservation areas, important prevention areas and key supervision areas for water and soil conservation, which play an important role in keeping the ecological balance of the basin and the whole island, natural disasters mitigation, and safeguarding the ecological safety of the whole island.

5.2.1 Delineation with Comprehensive Management of Land and Marine

Hainan Provincial Government firmly insisted on the principle of ecological priority and land-sea coordination, scientifically laid out and organized the implementation of the ERLs delineation, promoted the formation of a new pattern of ecological conservation, laid a solid ecological foundation for maintaining the ecological security of Hainan Free Trade Port, promoted the modernization of the ecosystem governance system and ability, sped up the construction of the National Ecological Civilization (Hainan) Pilot Zone.

In June 2015, Hainan Province was selected as the reform pilot for "integration of multi-plans". Hainan Provincial Government clearly regarded the ERLs as the spatial rigid constraint of "Integration of multi-plans" and required the urban construction, industrial development and infrastructure layout to follow the results of the ERLs' delineation to ensure the ecological safety of the province. Among them, Hainan Provincial Government focused on coordinating the relationship between land and marine, development and conservation. In the delineation of the "Three Zones & Three Lines", 27.4% of the land area of Hainan Island and 35.2% of the coastal management sea with important ecological functions and areas in need of mandatory and strict conservation have been classified into the ERLs, so as to enhance the ecological security barrier of Hainan Province.

Establishing a stable ecological conservation pattern on land and at sea. Based on the data of the Third National Land Survey, combined with the results of the "Dual Evaluation" of territorial space, Hainan Provincial Government has coordinated the integration and optimization of protected areas and adopted the latest coastline revision data. According to the space frame layout of the mountains and rivers, it clarified the ecological conservation pattern of Hainan Island and off-shore area. The space pattern is to take the central and southern mountains as the core, Songtao Lake and other important lakes as space nodes, and the Nandu He and other major river systems as well as the coastal belt serve as the ecological corridor. The important ecological spaces such as protected areas on the whole island and in coastal waters, centralized urban water source reserves, management areas of major rivers, lakes and reservoirs, important natural mountains and natural shorelines have been classified into the ERLs. The ERLs covered various typical ecosystems such as tropical rain forests, mangroves, beach wetlands, coral reefs and seagrass meadows, and built a comprehensive ecological conservation system of ecological green center—ecological corridor—ecological shoreline—ecological waters.

Maintaining and promoting regional biodiversity. Hainan Provincial Government has classified all the integrated and optimized protected areas into ERLs, and has delineated 7880.92 km^2 of the ERLs in the biodiversity conservation priority zone in the central and southern part of Hainan Island, accounting for 61.2% of the total area of this zone. Among them, the biodiversity conservation type of the ERLs has covered 5630.09 km^2, which has further improved the connectivity and integrity of Hainan tropical rainforest national parks, protected areas, nature parks and other protected areas and habitats of important species and enhanced ecological corridors. At the same time, strict conservation of delineated areas is of important significance for maintaining and improving the level of regional biodiversity and conserving important species.

Establishing an ecological barrier consisted of land and sea. With the newly revised coastline boundary, Hainan Provincial Government fully considered the integrity and natural connectivity of the coastal ecosystems to delineate the ERLs to the land side and the sea side respectively, so as to effectively connect the land and marine ERLs.

At the same time, areas with three typical tropical marine ecosystems of mangroves, coral reefs and seagrass meadows and 81.2% of offshore uninhabited islands were almost classified into the ERLs. The rate of spatial connectivity for land and coastal management sea reached 97%. Through overall planning on land and sea, Hainan Provincial Government effectively coordinated the conservation pattern on land and marine, promoted ecological co-construction and environmental co-governance in coastal zones (Fig. 5.2) and strengthened the construction of covering land and marine protected areas.

Fig. 5.2 National Marine Ecological Nature Reserve of Hainan Wanning Dazhou Island

Ensuring the safety of the living environment. Through delineating the ERLs, Hainan Provincial Government has handled the relationship between urban development and construction and the overall conservation of mountains, water, forests, farmlands, lakes, seas and islands and ensured that "The quality of Hainan's ecological environment can only be better, not worse". Meanwhile, Hainan Provincial Government straightened out the relationship between conservation and development, improved and enhanced ecosystem service functions, and built the ecological security pattern with complete structure and stable function to effectively ensure the safety of the living environment in Hainan Province.

5.2.2 *Hainan Tropical Rainforest National Park: Ecological Green Heart*

Hainan Provincial Government has taken the following measures to conserve the "ecological green heart":

First, it integrated and optimized the protected areas in Hainan Tropical Rainforest National Park. Hainan tropical rainforest is the most concentrated distribution, has the most diverse types, is the best preserved and is the largest contiguous continental island type rainforest in China. It is the representative of the island type tropical rainforest, the treasure house of tropical biodiversity and genetic resources and the ecological safety barrier of Hainan Island.

Hainan Tropical Rainforest National Park is one of the first five national parks published by the State Council in 2021, with a total area of 4269 km^2, including a core reserve of 2331 km^2. With the main mountains in the central mountainous area of Hainan Province as the skeleton, and with Wuzhi Shan and other mountains as the main body, Hainan Provincial Government has integrated the protected areas and natural parks and extended them to surrounding natural forests and ecological national non-commercial forests, and established a nature reserve system with national parks as the main body and coordinated the conservation and development of the central ecological conservation area.

Second, Hainan Provincial Government carried out the ecological relocation. Based on the conservation of the authenticity and integrity of the rainforest ecosystem, Hainan Provincial Government has formulated an ecological relocation plan to relocate 11 natural villages located in the core reserve of Dongfang, Wuzhishan, Baoting and Baisha counties (cities), managed the collective land by conversion, lease, easement and contract, so as to systematically conserve national parks, form a new pattern of harmonious coexistence between man and nature, and promote the construction of national Ecological Civilization pilot zones. The relocation of Hainan ecological migration has fulfilled the commitment of moving into new house, keeping long livelihood, getting richer and achieving a win-win status of regional ecological restoration for the villagers, enable villagers to adapt to ecology more consciously, to love ecology and to conserve ecology by conserving Lucid Water and Lush Mountains and harvesting Golden and Silver Mountain, so as to achieve mutual promotion of ecology, production and life.

5.2.3 *Implementing Restoration to Build Ecological Security System*

Hainan Provincial Government has strengthened ecological restoration and established a territorial space conservation pattern for mountains, rivers, forests, farmlands, lakes and islands.

First, strengthening the conservation and restoration of mangroves. With a good reputation as "coastal guard" and "marine forest", the mangrove forest is a special wetland ecosystem in the transitional zone between the land ecosystem and the marine ecosystem, which plays an important role in purifying sea water, withstanding wind and waves, conserving and improving the ecological environment, maintaining biodiversity and ecological security in coastal areas. For this purpose, Hainan Provincial Government has strengthened the conservation and restoration of mangroves. For example, the Overall Plan of Huachangwan Mangrove Nature Reserve has been compiled, and the area has been classified into the ERLs. The conservation scope and functional zoning have been defined, the damaged landform has been restored, mangroves have been replanted and the natural environment has been gradually restored, and the destroyed and lost "sea forest" has reappeared (Fig. 5.3).

Second, carrying out ecological restoration of coral reefs. Coral reef is known as the "tropical rainforest" of the sea floor, which is one of the most biologically diverse ecosystems on Earth. In recent years, due to tourism development, the coastal management sea of Sanya City has been eroded and the coral reef on the sea floor has been destroyed. To restore the submarine ecosystem, Sanya municipal government has successively placed artificial reefs in the nearby waters such as Phoenix Island and Luhuitou, carried out "forest planting" at the sea floor and transplanted coral seedlings suitable for the sea area to gradually restore the coral ecosystems (Fig. 5.4).

Third, carrying out a 3-year act for ecological conservation and restoration. By the end of 2022, 126.59 km of shoreline in Hainan Province has been cleaned up and restored. The marine ecological environment in the province has been significantly improved through clearing aquaculture ponds and tidal flats to make way for coastal wetlands such as mangroves. At the same time, an "increase green and conserve blue" ecological security system has been built. It includes the "central water tower", the ecological corridor consisting of 3 rivers and 13 river basins, more than 63% of the natural shoreline, 100 hm^2 of "coral reef areas" has been built through a 3-year act of ecological restoration in the south-central mountainous areas and the northern plain hilly areas of Hainan Island, as well as the ecological restoration of the coastal zone and coral reefs.

Fig. 5.3 Corner of Huachangwan Mangrove Nature Reserve after ecological restoration

Fig. 5.4 Coral reef ecological restoration in Wuzhizhou Island sea area, Sanya

5.2.4 Strengthening Supervision to Improve the Development and Conservation of Territory

Hainan Provincial Government has taken many measures to strengthen the supervision of the ERLs.

Formulating rules and regulations and strengthening supervision. First, Hainan Provincial Government took the lead in issuing the regulations on the management of the ERLs in Hainan Province, clarifying the scope of the ERLs, the responsibilities of the government and relevant departments, entry conditions, regional control, adjustment conditions, responsibility investigation, etc., and implemented the strictest ecological and environmental conservation. Second, Hainan Provincial Government issued regulations on the conservation and development of the coastal zone, and implementation rules of coastal zone conservation and development, and strengthened environmental resources and conservation of coastal zones and standardized the development and utilization management of coastal zones. Third, according to the notice on strengthening the management of ERLs issued by MNR, MEE and NFGA, the current situation in Hainan Province, catalogue of the ERLs access management in Hainan Province has been revised, and the positive access list and the ERLs' control rules have been clarified, so as to further strengthen the conservation of ecological space and improve the economical and intensive use of territorial spaces.

Carrying out special inspections and establishing a monitoring system. First, Hainan Provincial Government carried out special inspections of coastal zone conservation and development and special supervision of the ERLs across the province to play the role of overall planning and supervision. Second, Hainan Provincial Government established a three-dimensional monitoring system of "sky inspection, ground patrol and network reporting" by using satellite remote sensing monitoring and video surveillance technology to timely discover and effectively contain all kinds of illegal activities that damage natural resources of the ERLs and coastal zones.

Setting up an information platform and an early warning and monitoring mechanism. First, Hainan Provincial Government set up the provincial comprehensive management platform for "multi-plan integration" information and established a whole-domain, all-factors, whole-process, digital territorial space governance system to realize "machine management planning". Second, Hainan Provincial Government developed the "one blueprint" public version and App (application software), completed the Web side (computer side web version), mobile App and WeChat mini program, embedded with "service on code" and "citizen cloud in coconut city" so that the public can inquire about the delineating and planning of various controlled lines such as "Three Zones & Three Lines". Third, Hainan Provincial Government established a real-time early warning and monitoring mechanism of the ERLs, explored setting electronic fences to remind of prohibited behaviors of personnel entering the ERLs through the real-time mobile signaling data of the three major operators of China Mobile, China Unicom and China

Telecom, so as to conduct real-time monitoring of human activity intensity within the ERLs.

5.3 Heilongjiang Province: The Northeast Forest Belt[4]

Heilongjiang Province is a natural treasure house of biodiversity in China, which is located in the northeast of China, with rich ecological resources, rare and endangered animals and plants such as the amur tiger, the red-crowned crane, the northeast taxus chinensis and tilia amurensis. As an important ecological function area in China, the province has four important ecological function areas, namely Da Hinggan Ling and Xiao Hinggan Ling, Changbai Shan Mountain, Sanjiang Plain and Songnen Plain, which are important parts of the northeast forest belt and play important roles in maintaining ecological balance across the region, alleviating natural disasters and ensuring ecological security.

5.3.1 Delineation Practices

Heilongjiang Provincial Government insisted on the principles of "giving priority to conservation, delineating seriously; scientific and objective, seeking truth from facts; upholding the bottom line, strict conservation; strict requirements, collaborative promotion; adjusting measures to local conditions, overall optimization" to carry out the ERLs delineation (Fig. 5.5).

First, promoting by high level government planning in advance. Heilongjiang Provincial Government attached great importance to the ERL delineation, organized and held high-level work meetings on the ERL delineation work of the province.

Second, regional coordination, overall optimization. Heilongjiang Provincial Government focused on the connection and in-depth docking of the ERLs with neighboring provinces, gave consideration to the demands of different regions for the conservation and development of territorial space, and further optimized the structure and layout of cross-regional ERLs, to ensure that the ERLs are scientific, authoritative, consistent and performable.

[4] This part was written by Yu Hongfeng, Zhang Zhong, Zheng Hao and Xu Enyu. Yu Hongfeng, the director of the Territorial and Spatial Planning Bureau, the Department of Natural Resources of Heilongjiang Province; Zhang Zhong, the deputy director of the Territorial and Spatial Planning Bureau, Department of Natural Resources of Heilongjiang Province; Zheng Hao, the vice president of Heilongjiang Provincial Territorial and Spatial Planning Institute; Xu Enyu, the chief of the Territorial and Spatial Special Planning Institute, Heilongjiang Provincial Territorial and Spatial Planning Institute.

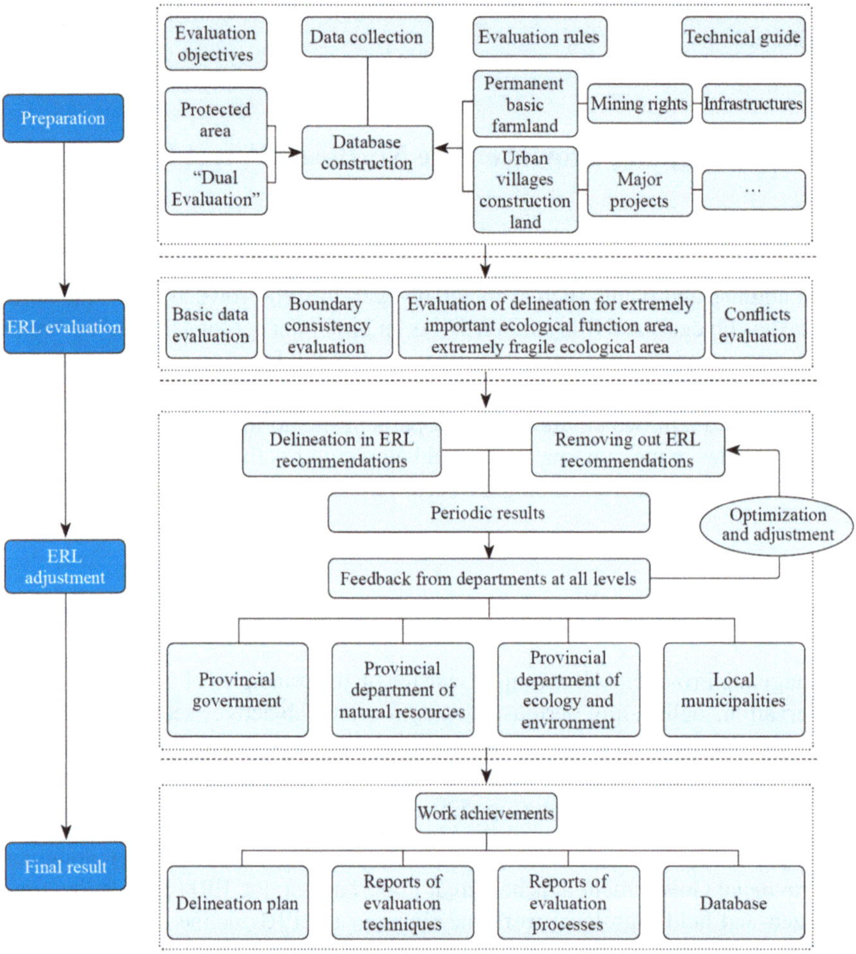

Fig. 5.5 Technology flowchart for ERL evaluation and adjustment

Third, scientific delineation, comprehensive conservation. Heilongjiang Provincial Government scientifically carried out the "Dual Evaluation" to identify the extremely important ecological areas and extremely fragile ecological areas, then incorporated them into the ERLs together with other areas that need to be strictly conserved, to achieve the overall conservation of important ecological spaces.

Fourth, unifying base data, strengthening cohesion. Heilongjiang Provincial Government summarized and checked the basic data required for the ERL evaluation and adjustment, confirmed the legal compliance of data sources and converted coordinate parameters to ensure the reliability of basic data. At the same time, the government fully linked with the integration and optimization of protected areas as the core space composition of ERLs.

Fifth, finding out the contradictions and handling them later. In the process of adjustment, Heilongjiang Provincial Government always insisted on the working method of "finding out the contradictions first, handling them later", and properly dealt with the problems left over from history, according to the principle of "properly handling historical contradictions and reducing unnecessary new contradictions", so as to realize scientific, reasonable and operational handling of contradictions & conflicts.

Last, people-oriented and quality-improving focus. In the process of the ERLs delineation, Heilongjiang Provincial Government focused on the relationship among agricultural production, urban development and construction and the overall conservation of mountains, rivers, forests, farmlands, lakes, grasslands and sand, and balanced the relationship between overall conservation and development.

5.3.2 Delineation Achievements

Heilongjiang Provincial Government has integrated and optimized the extremely important ecological function areas, such as protected areas, water conservation, biodiversity conservation, soil erosion control, windbreak and sand fixation etc., and the extremely fragile ecological areas, such as soil erosion areas, land desertification areas, etc., as well as the areas with potentially important ecological values that must be strictly conserved, which have been included in the ERLs. Mining rights, permanent basic farmland, towns and villages, artificial commercial forests, major construction projects and other types of complex contradictions & conflicts have been handled properly to optimize and adjust the ERLs.

Establishing a solid ecological security barrier. The ERLs in Heilongjiang Province covered 152,700 km^2, accounting for 32.44% of the total area of the province. From the perspective of distribution, the ERLs were basically consistent with the ecological security pattern of "three mountains, four rivers and multiple scenic spots" in Heilongjiang Province, mainly distributed in the forest ecological function areas (Da Hinggan Ling and Xiao Hinggan Ling) and the Changbai Shan (Zhangguangcai Ling-Laoye Ling-Wanda Shan), and the ecological space consisting of the Heilong Jiang, Ussuri Jiang, Songhua Jiang, Nen Jiang and nearby important wetlands. With the ecological conservation pattern as the guidance, Heilongjiang Provincial Government coordinated the overall conservation, systematic restoration and comprehensive management of ecological space, improved the quality of ecological space, and built a strong ecological security barrier in the northern territory.

Overall coverage of key species and habitats. The ERLs in the province were concentrated in the Changbai Shan, Da Khingan Ling and Xiao Khingan Ling, Sanjiang Plain and other important ecological function areas, covering forests, grasslands, wetlands, rivers, and the province's biodiversity distribution of key areas, which can conserve rare and endangered species and their habitats to the maximum extent. For example, Amur tiger and Leopard National Park, jointly designated by Heilongjiang Province and Jilin Province, is the historical natural distribution area of amur tigers and amur leopards in China and the only area with wild settled populations and breeding families. It is one of the most important and most biodiverse areas in China's natural ecosystem. Its establishment has important significance for conserving the habitat and reproduction of wild populations of Amur tiger and Amur leopard, keeping the authenticity and integrity of the ecosystem, and promoting cross-border conservation cooperation of rare and endangered species.

In addition, according to the coverage of key species, the Heilongjiang Central Station Black-billed Grouse National Nature Reserve and Heilongjiang Bishui Chinese Merganser National Nature Reserve as protected areas covering first-class national key wildlife conservation to conserve rare and endangered animals have been delineated. The country's largest dinosaur fossil, black bee and natural forest reserve of ERLs have been delineated, so as to achieve full coverage of critical endangered species and habitats.

Enhancing the conservation and control of ecological resources. Heilongjiang Provincial Government strictly conserved and controlled the use of ecological resources such as water resources, forest lands, grasslands and wetlands in Heilongjiang Province as follows:

First, strictly conserving and controlling the forest. Heilongjiang Provincial Government implemented graded forest conservation and management, and strictly implemented the management system of limited logging with quotas and certification. Heilongjiang Provincial Government comprehensively conserved natural forests in the province, completely stopped commercial logging of natural forests, did not allow transforming natural forests into artificial forests, did not allow transforming non-commercial forests into commercial forests, carried out managing natural forest conservation and non-commercial forest simultaneously, and promoted the rehabilitation of natural non-commercial forests which can be cultivated, updated and cut with low quality and low efficiency, but the trees in protected areas were not allowed to be cut. Heilongjiang Provincial Government strictly controlled the logging area of commercial forests, and implemented simultaneous logging and breeding plans.

Second, strictly conserving and controlling the use of basic grasslands. Heilongjiang Provincial Government strictly conserved natural grasslands in important ecological areas such as the sources of major rivers, and prohibited unauthorized changes in the use and nature of grasslands, and the development and utilization of activities that do not conform to the functional orientation of grassland conservation. Heilongjiang Provincial Government scientifically delimited basic grasslands, strictly conserved and managed them, implemented a system of grazing prohibition and rest grazing, and balanced between forage and animals. Heilongjiang Provincial Government strengthened ecological restoration and management of degraded grasslands and improved the quality and functions of grasslands, and developed modern grass industry, grassland ecological animal husbandry and grassland ecological tourism, according to local conditions.

Third, wetland conservation and usage controlling. Heilongjiang Provincial Government implemented wetland system conservation and graded management, and carried out the restoration of wetland vegetation, restoration of topography and landform, maintained natural wetland shoreline, and connected rivers and lakes in the national important wetlands, national wetland parks with important ecological locations, wetland protected areas above the provincial level (including provincial level). The government strengthened the conservation of wetland and rare migratory birds' migration and breeding areas, steadily promoted the return of cultivated land to wetland, restored the degraded wetlands, and improved the functional stability of regional wetland ecosystems. The government controlled the total area of wetlands and strictly controlled the occupation of important wetlands by construction projects. Under the premise of effective conservation of wetland resources, Heilongjiang Provincial Government properly and reasonably developed the education, ecological tourism, ecological green planting and breeding industries of nature wetland.

Strengthening regional ecological restoration. With the ERLs as the core, Heilongjiang Provincial Government has promoted ecological restoration across the region. Based on relatively complete natural units such as mountains, rivers and lakes, comprehensively considering the systematic and integrity of natural ecology, making coordinated efforts to promote the systematic management of mountains, rivers, forests, farmlands, lakes, grasslands and sands, an overall pattern of "one belt, two areas and four corridors"(one belt: the Northeast forest belt dominated by "three mountains"; two areas: Songnen Plain and Sanjing Plain; four corridors: four river corridors, namely, Heilong Jiang, Songhua Jiang, Nen Jiang, Ussuri Jiang) for the conservation and restoration of territorial spaces was formed. According to the natural geographical spatial pattern of the province and the distribution of the ERLs, Heilongjiang Provincial Government delimited the Da Hinggan Ling permafrost and forest ecological conservation and restoration area, Xiao Hinggan Ling forest and mine ecological conservation and restoration area, Changbai Shan (Zhangguangcai Ling-Laoye Ling-Wanda Shan) forest and mine ecological conservation and restoration area, Songnen Plain as the black soil resources reservation, agriculture, comprehensive ecological conservation and restoration area of forest and animal husbandry wetland. It delineated the ecological conservation and restoration area of Sanjiang Plain as black soil resources and farmland wetland reservation, the restoration area of the border ecological conservation belt. It strengthened the conservation and management of natural forests, natural grasslands, natural wetlands and other ecological resources according to the law, and improved the cultivation of reserve resources to further reverse the trend of succession. The government constantly promoted water purification and vegetation restoration, and strengthened ecological functions such as water conservation, climate regulation, and species conservation. The government steadily promoted the conservation and restoration of important habitats and construction of ecological corridors, and strengthened biodiversity conservation.

5.3.3 Implementation of Territorial and Spatial Planning in Cities and Counties

Focusing on guidance and supervision during the delineation process. In the process of delineation and adjustment, Heilongjiang Provincial Government coordinated the relationship between speed and quality, development and conservation, and the parts and the whole in the face of the contradictions between overall development and conservation in various regions and the difficulties in implementation of control indicators. It promoted the delineation and adjustment of the ERL in 13 cities across the province through measures as follows:

At the beginning of the work, a leading group for the delineation work was established, with a comprehensive coordination group, a technical guidance group and four district guidance working groups, so as to implement district guidance and full supervision in each place. The meetings at the provincial level have been held to report the results of the delineation of the ERLs for advance scheduling, and timely grasped the modification and improvement of the delineation results across the province. In view of the problems existing in various places, on-site questions were answered. Technical guidelines such as work plan for review of the results of the ERLs and requirements for the result data collection, etc., were timely issued to scientifically guide the delineation in all cities, considered both conservation and development, and ensured that the province's work was completed with high quality.

Delineation results were consistent with the ERL layout. The ERLs have been implemented in 13 cities of Heilongjiang Province as the basis for land use approval. Among them, Da Hinggan Ling region, Yichun City and Mudanjiang City had a higher proportion of the ERLs, accounting for 74.40%, 46.40% and 43.43% of the total land space, respectively. Prefecture-level cities with less than 10% of the territorial space, including Qitaihe City and Qiqihar City, accounting for 8.29% and 9.06%, respectively. The cities with 10% to 30% included Kiamusze City, Daqing City, Jixi City, Shuangyashan City, Hegang City, Heihe City and Suihua City.

5.3.4 Management Practices

The provincial government safeguarded the seriousness and authority of ERLs earnestly through the measures as follows:

First, arbitrarily adjusting the scope of the ERL is forbidden. The ERLs should not be arbitrarily adjusted once the results of the delineation are approved. Local governments should regard the ERLs as an important basis and precondition for comprehensive decision-making, and various construction projects should give priority to avoiding ERLs; if impossible, they must meet the requirements for the management of ERL, and resolutely uphold the seriousness and authority of ERL.

Second, the ERLs should be included in the approval matters for land construction. In the process of approving land construction, the ERLs should be regarded as the mandatory review content, strictly review the use, layout, compliance, necessity and other contents of construction projects, and implement "one vote veto" for construction projects that involve the ERLs but do not meet the control requirements.

Third, establishing a joint examination mechanism for the approval of construction projects within the ERLs. In the construction land pre-examination and site selection, land application for approval, the provincial nature resource authority should solicit opinions from provincial ecological environment and forestry departments, ask them to participate in the relevant demonstration review, thus putting forward clear opinions on whether the construction project meets the requirements of protected areas, forest occupation, drinking water source, ecological environment conservation, etc. For a construction project which does not conform to the relevant laws and regulations, land application for approval should not be accepted.

The provincial government firmly conducted the identification of human activities within the ERLs through the following measures:

First, strengthening the necessity for site selection. Heilongjiang Provincial Government required that all localities and departments should give priority to avoiding the ERLs in the process of selecting sites for construction projects. For those that meet the control requirements and cannot avoid the ERLs, construction project units or industry authorities should strengthen the necessity of project construction, the rationality of site selection, and the comparison of several options for site selection, and explain the inescapability. Only the projects that meet the requirements can go through the admission procedures.

Second, strengthening classified examination and approval. If human activities are involved in the new construction land, the government at county level should apply to the provincial government to issue a determination opinion. Among them, for the national major projects that really need to occupy the ERLs, the provincial department of natural resources management should organize to demonstrate, apply to the provincial government for the issuance of the determination opinion on the inescapability of ERL, and report to the State Council for approval of land use.

Third, strengthening work responsibility. When handling the determination opinion, the municipal government should issue the preliminary determination opinion on allowing limited human activities within ERLs. In the construction project land pre-examination and site selection, the application for agricultural land conversion and land expropriation are submitted along with the determination opinion. If national parks, nature reserves, tourist attractions and other protected areas, important wetlands, drinking water source reserves and other areas are involved, the relevant competent departments should give a definite opinion step by step.

The provincial government adopted information technology to improve the management efficiency of ERLs.

First, the ERLs were included in the "one map" system of territorial and spatial planning. Heilongjiang Provincial Government has incorporated the scope of the ERLs and relevant information into the "one map" implementation supervision information system for territorial and spatial planning, and connected with the approval system of construction land, review of development plans, and certification of mining rights, so that all departments can rapidly check the compatibility with the ERLs.

Second, speeding up information sharing. Heilongjiang Provincial Government has distributed the implementation supervision information system account of territorial and spatial planning "one map" so that relevant departments can invoke the scope and information of ERLs and provide services in planning, layout, project site selection and other decisions, and give priority to avoiding the ERLs.

5.3.5 Thoughts on Strengthening the Management System

Promoting the "legalization" of the ERL supervision. Robust administrative regulations are the basis for strictly following the ERLs. Covering a wide area, involving many parties and complex interests of all parties, the ERLs need clearer and complete legal safeguards in the conservation responsibility, control measures, responsibility investigation, illegal investigation, etc. At present, relevant laws and regulations have been issued and implemented for protected areas, tourist attractions, geological parks, and drinking water sources, and "Law on Nature Reserves" and "Law on National Parks" have also been researched and formulated, but the management of the ERLs is only issued in the form of policy documents. Hence, it is suggested to issue laws and regulations on the management of ERL as soon as possible to meet the legal needs of the ERL management, clarify the main body of responsibility for the ERL supervision, straighten out the relationship between the ERL management and relevant regional laws and regulations, clarify control rules for the ERL, clarify the law enforcement regulations on the violation of ERL, and ensure that the ERLs are governed by laws.

Promoting the "scientific modernization" of the ERL supervision. An accurate and efficient monitoring and conservation effectiveness evaluation system is the technical support for strictly following the ERLs. Carrying out the ERL supervision, and establishing a long-term mechanism for the ERL management and control are the important links of implementing Ecological Civilization, and an important part of MNR to promote green and low-carbon development. For the ERL monitoring and supervision of Heilongjiang Province in the future, it is necessary to improve the technical system for the ERL evaluation and monitoring, thus formulating long-term evaluation, monitoring indicator systems, technical methods for ecological environment quality and human activities within the ERLs. Establishing a "monitoring-evaluation-early warning" mechanism will help management departments with scientific decisions.

Based on the "one map" of territorial and spatial planning, combined with multiple sources of information such as field patrol, monitoring and big data inside and outside protected areas, as well as the ecological environment supervision platform data of the ecological environment department, a multi-level monitoring and conservation effectiveness evaluation technical system (Fig. 5.6) has been established through comprehensively adopting remote sensing satellites, aerial drones, field survey and ground observation, and relying on the "sky-space-earth-net" integrated natural resources investigation and monitoring technology system.

Promoting the "convenience" of the ERL information. In the future, Heilongjiang Provincial Government should carry out boundary surveys and setting up signs, establish boundary posts and sign boards in areas with frequent human activities, enhance the timeliness of warning alerts through electronic fencing, and set up a convenient work platform that can be visualized, queried and located within the ERLs, so as to improve the convenience of public inquiry and avoidance of the ERLs, and strengthen the real-time conservation of ERL. At the same time, the ERLs will be included in the province's "Investigation Cloud" platform in time to

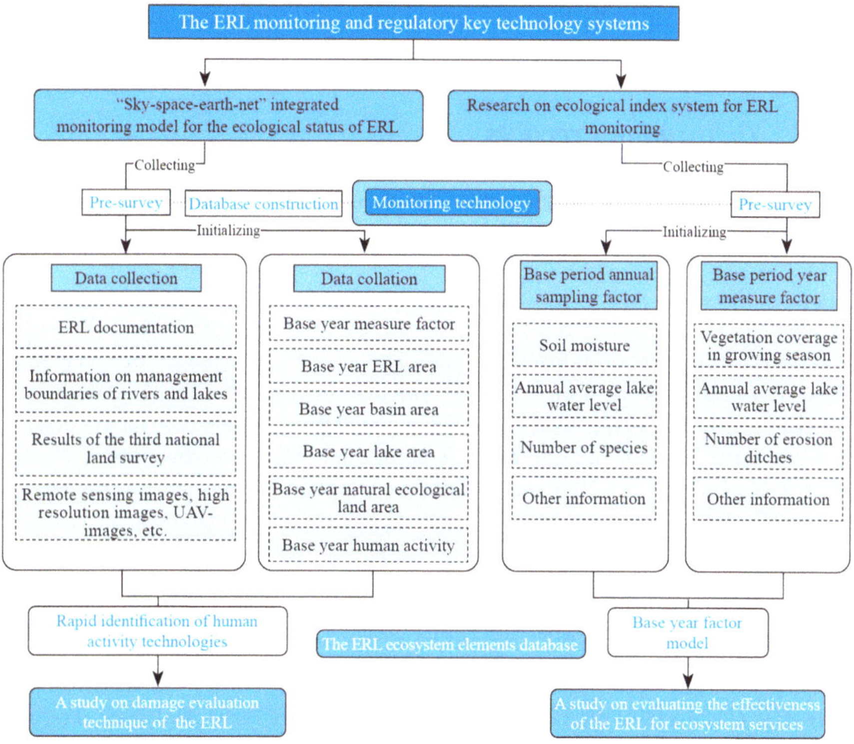

Fig. 5.6 Logic diagram of the ERL monitoring and supervision system

promote data sharing in various departments and industries in the province. By increasing the disclosure of monitoring information at all levels for the ERLs delineation, it is necessary to give full play to the subjective initiative of the public on the subject of ecological conservation rights and responsibilities, avoid the ERLs at the source, and reduce unnecessary adjustments.

5.4 Zhejiang Province: Transforming Resources into Valuable Assets[5]

Zhejiang Province is located on the southeast coast of China and the south flank of the Chang Jiang Delta. To safeguard and maintain ecological security, uphold the bottom line and lifelines, the total area of ERLs in Zhejiang Province is 36,800 km^2, accounting for 24.59% of the land area of Zhejiang Province. Among them, the land ERLs cover 22,200 km^2, accounting for 21.02% of the total land area of Zhejiang Province; the marine ERLs cover 14,600 km^2, accounting for 33.15% of the maritime area under Zhejiang Province's jurisdiction; 2732 uninhabited islands have been classified into ERLs. As the first ecological province in China, Zhejiang Province constantly promotes the construction of a Beautiful Zhejiang and Ecological Civilization through delineating and strictly enforcing the ERLs.

5.4.1 Delineation Practices

5.4.1.1 Enhancing the Implementation of National Ecological Responsibilities

Zhejiang Province, rich in biodiversity, is the main distribution area of the national "Huangshan-Huaiyushan Biodiversity Conservation Priority Area" and "Wuyishan Biodiversity Conservation Priority Area". The area of the ERLs in the two priority areas accounts for 40% of the total ERLs area in the province, 16% higher than the provincial average, respectively. The ERLs in the province are distributed in a pattern of "three barriers and one belt". Among them, the "three barriers" including the ERLs of southwest Zhejiang mountain ecological barrier, northwest Zhejiang

[5]This part was written by Gu Hao, Tang Xiaolu, Jiang Yanhong, Xu Hongbo and Mao Wanliu. Gu Hao, the vice director of the Department of Natural Resouces of Zhejiang Province; Tang Xiaolu, the vice director of the Territerial and Spatial planning Bureau, the Department of Natural Resources of Zhejiang Province; Jiang Yanhong, a four-class consultant of Territerial and Spatial planning Bureau, the Department of Natural Resouces of Zhejiang Province; Xu Hongbo, a senior engineer of Zhejiang Provincial Mapping Scientific Institute; Mao Wanliu, an engineer of Zhejiang Provincial Mapping Scientific Institute.

mountain ecological barrier and middle east Zhejiang mountain ecological barrier cover nearly 13,000 km^2, accounting for nearly 2/3 of the province's land ERLs. In the ecological space of the Changjiang Economic Belt, the Tianmushan-Huaiyushan forest area and the hilly and mountainous forest area in southeast Zhejiang Province have been classified into ERLs of nearly 14,000 km^2, accounting for more than 2/3 of the province's land ERLs. The Zhejiang region of the national "Priority area for Biodiversity Conservation in the East China Sea and Taiwan Strait" is consistent with the coastal area of east Zhejiang in the "three barriers and one belt". Among them, "one belt" areas include important estuaries, important coastal wetlands, mangroves, mainland coastline conservation areas, islands and surrounding seas, the marine ERLs with an area of 7467 km^2, accounting for more than 1/2 of the province's marine ERLs. An ecological barrier has been formed from north to south to improve the implementation of national ecological responsibilities.

5.4.1.2 Carrying Out Evaluations on Ecological Conservation Importance

To effectively support the ERLs' delineation, Zhejiang Provincial Government has scientifically analyzed the province's resource endowments and environmental conditions according to the national standards and the reality, and carried out the evaluation of the ecological conservation importance of land and sea areas.

1. Grading evaluation of the importance of land ecological conservation. The evaluation of the importance of land ecological conservation in Zhejiang Province included the evaluation of the importance of ecosystem functions, such as biodiversity conservation, water conservation, water and soil conservation, and the evaluation of ecological vulnerability, such as water and soil erosion, as well as the evaluation of the comprehensive formation of extremely important ecological conservation areas. The indexes of the evaluation results were calculated according to the guideline on the suitability evaluation of resource and environment carrying capacity and territorial spatial development, and the guideline on the ERLs delineation. From a national perspective, Zhejiang Province has a good natural ecological background, and its richness of various resources, such as forests, wetlands, animals and plants. The volume of those natural resources is higher than the national average level. Therefore, stricter evaluation thresholds and grading were used for the ecological conservation importance evaluation of Zhejiang Province, and the top 20% of the thresholds were classified into five importance grades to identify the ecological resources required for conservation. The evaluation object has been checked and inspected on site time and again, and the thresholds determined were more in line with the actual conditions of Zhejiang Province. The evaluation results not only met the scientific evaluation method, but also support the ERLs' delineation.

2. Optimizing marine ecological conservation importance evaluation indexes. Based on the guide for technical evaluation of the importance of marine ecological conservation issued by the MNR, according to the marine reality of Zhejiang Province, the evaluation index system of mainland coastline and uninhabited islands in the guide was optimized, and the importance of marine ecological conservation was evaluated from the aspects of sea area, mainland coastline and uninhabited islands. In terms of marine areas, the marine extremely important areas of ERLs, including coastal estuaries, bays, coastal wetlands, mangroves, protected areas, and spawning site reserves, were identified from the species level and the ecosystem level. In terms of continental shorelines, the evaluation of ecological conservation importance for the shorelines supported by protected areas, important coastal wetlands and mangroves was carried out, in the way of "line-surface" combination, on the basis of ecological service function, ecological sensitivity and conservation function. In the context of uninhabited islands, a quantitative evaluation of the ecological function importance was carried out. This evaluation considered three aspects: island vegetation, island intertidal benthos community, seabird migration routes, and other habitat conservation functions; territorial sea rights functions; and sensitivity to coastal erosion, sand loss, etc. This evaluation aimed to identify significant islands for ecological conservation. This evaluation has subsequently served as a crucial foundation for the delineation of marine and uninhabited islands within the ERLs' framework.

5.4.1.3 Partitioning Uninhabited Islands with Ecological Function

Zhejiang Province has the largest number of islands among the coastal provinces, and most of them are uninhabited islands. Delineation of the ERLs for uninhabited islands has an important significance for maintaining the ecological security and safety of its islands, and promoting the reasonable use of island resources.

According to the evaluation results of the ecological conservation importance of uninhabited islands in the province, and the qualitative analysis results of island area, island distance from the shore, and development and utilization status, the ecologically important islands in the province have been determined. These islands include islands for territorial waters and national defense purposes, uninhabited islands with special conservation values, such as rare and endangered wildlife and habitats, and important natural relics. At the same time, in view of the connectivity and integrity of the island and the surrounding sea ecosystems, the coastlines of ecologically important islands and their surrounding sea areas were classified into the ERLs. Due to the large quantity and wide distribution of uninhabited islands in Zhejiang Province, the ERLs of ecologically important islands were delineated, based on the evaluation conducted from the perspective of important ecological functions.

5.4.1.4 Innovation in Development of Quality Control Software

To achieve the standardization inspection of the organization and format of the results of the ERL delineation in cities and counties, the ERL quality control software was developed, which can automatically analyze and review the proportion of extremely important ecological conservation areas, national non-commercial forests, protected areas, drinking water source reserves and other areas that should be classified into the ERLs of city (county), and can automatically analyze and distinguish the transfer situation of cultivated land, artificial commercial forests, urban and other human activity areas within the ERLs. The software was distributed to all cities (counties) for self-inspection, so as to improve the overall review efficiency of the ERLs in the province. In the quality control software, the inspection rules for the fragmentation degree of map polygon and "skylight" are innovatively set, and the fragmentation degree of ERL is reviewed with the county (city) as a unit.

First, the fragmentation situation of ERL was checked according to the fragmentation index of the map patch. The map patch fragmentation of the land ERLs should not exceed 0.2 pcs/km^2, and the index can be appropriately adjusted in plain areas and island areas.

Second, the fragmentation situation of "skylight" was checked according to the fragmentation index of "skylight". The fragmentation degree of the "skylight" in the land ERLs should not exceed 0.3 pcs/km^2, and the number of "skylight" should be reduced in the county (city). The inspection of ERL fragmentation was carried out by using quality control software to ensure that the map patches of ERL are concentrated and contiguous.

5.4.1.5 Strengthening the Organization of Work

MNR and MEE jointly issued a document to deploy the evaluation and adjustment of the ERLs in June 2019. The Department of Natural Resources of Zhejiang Province, together with the Department of Ecology and Environment of Zhejiang Province and the Forestry Bureau of Zhejiang Province, adopted the organizational mode of "provincial guidance, city and county execution" and the working mode of "up twice, down twice". They insisted on the basic principles of "conservation priority, overall stability, contradiction resolving and overall planning" under the overall thought of "multi-plan integration" and "combination of delineation and management", and comprehensively promoted the evaluation and adjustment of ERL.

1. Deploying comprehensively, establishing committees and examining strictly. The Department of Natural Resources of Zhejiang Province took the lead in the delineation work, and deployed to promote the overall work together with the Department of Ecology and Environment of Zhejiang Province and the Forestry Bureau of Zhejiang Province. The provincial department of natural resources established a working group together with six provincial technical support units,

which were responsible for the technical and administrative review of the results of ERL initially delineated in cities (counties), and formulated review opinions to ensure the quality of the results.

2. Strengthening cooperation, coordination at different levels, and departmental cohesion. First, the provincial government communicated with MNR and neighboring provinces such as Fujian Province, Shanghai Municipality, to solve the overlapping problems of the marine ERLs crossing Zhejiang and Fujian, Zhejiang and Shanghai. Second, it set up a working mechanism for departmental cooperation. Delineation of the ERLs has fully docked with the provincial Department of Ecology and Environment and the provincial Department of Transportation. The ERLs has docked with the provincial Territorial and Spatial Planning, the ChangJiang Economic Belt Territorial and Spatial Planning, and the Demonstration Zone Planning of Chang Jiang Delta Ecological Green Integrated Development to ensure that important ecological areas in the planning are classified into the ERLs. Third, it set up the linkage system of the province, municipalities and counties, unified technical standards and formulate review methods. The governments of cities (counties) attached great importance to it, and the results of the evaluation and adjustment of ERL were reported to the provincial department of natural resources after the approval of the municipal government (Fig. 5.7).

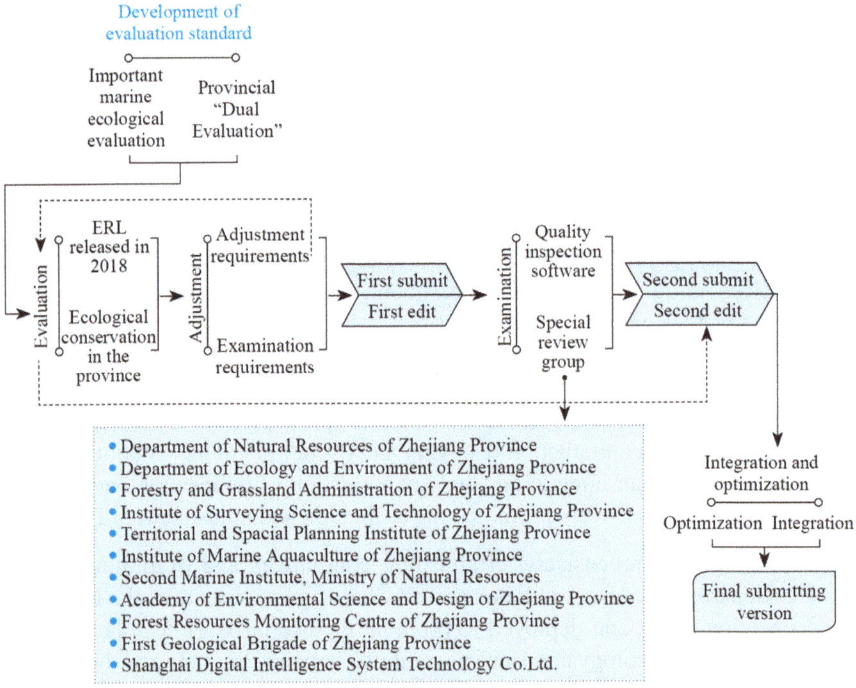

Fig. 5.7 Flowchart of ERL evaluation and adjustment

5.4.2 Issuing the Implementation Guideline

In December 2022, Zhejiang Provincial Government issued the implementation opinions on strengthening the supervision of ERL. This is the first normative document for ERL supervision issued and implemented by the provincial government after the demarcation of the "Three Zones and Three Lines". The provincial government actively innovated according to the national ERL management requirements, did a good job in the ERLs supervision, and strived to form a new pattern of ecological conservation in which local governments attach importance to it, all sectors pay attention to it, and all people participate in it.

1. Regarding "planning" as the precondition, coordinating the conservation and utilization, and promoting green production and lifestyle. By means of "control", it stipulated the rules for ecological resources conservation and utilization, the requirements for approval, and the management requirements for other limited human activities. Using geo-informatics technique to construct the ERLs supervision application scenarios, it built a data sharing, business linkage, collaborative management, and unified and efficient ERL supervision system among the "provincial-municipality-county". With "compensation" as the driving force, it gradually set up an ecological conservation compensation mechanism with strong government leadership, orderly social participation, and effective market regulation.
2. Clarifying the specific types and approval requirements for allowing limited human activities. The government clarified the specific types of linear infrastructure in line with the territorial and spatial planning at or above the county level, production and living facilities for indigenous residents and other legitimate rights and interests subjects, and the construction and maintenance of necessary public facilities. At the same time, it stipulated strictly the approval of construction, the land use and planning permission, clarified the approval requirements for the use of sea, island and forest, so as to reflect the features of high quality ecological space conservation in Zhejiang Province.
3. Extending the working chain of human activities supervision. First, it strengthened the guidance and control of territorial and spatial planning, and put forward the requirements of guidance behaviors and ecological restoration for special plans and detailed plans involving the ERLs. Second, it promoted green production ways and lifestyles, controlling the intensity of ecological green agriculture, mineral exploration and mining, linear infrastructure construction, traffic intensity, etc. Third, it properly handled the legacy of history and the future ecological conservation compensation, let the local governments give full play to the leading role to ensure ecological security and social stability.
4. Strengthening the supervision mechanism of the ERLs. First, it clarified the main responsibilities of local governments and the departments that involve the departments in charge of development and reform, finance, water conservancy, agriculture and rural areas, in addition to natural resources, ecological environment, forestry authorities. Second, it insisted on the lead of digital reform,

depending on the provincial spatial governance digital platform to develop the ERLs supervision application scenarios, promoting the construction of a unified and efficient supervision platform of the ERLs with the function of data sharing, business linkage, management coordination inner the levels of "ministry-province-municipality-county", and improving the digital supervision level of ERL.

5.4.3 *Continuously Improving the Supervision and Monitoring*

1. Strict approval for the land and marine use of ERL. The Implementation Opinions on Strengthening the Supervision of the ERLs stipulated that for projects involving the land and marine within the ERLs, the demonstration of the inescapability of the ERL's should be conducted. To ensure the compliance of limited human activities within the ERLs, as well as the legality, rationality, science and operability of construction projects, Zhejiang Provincial Government has issued a series of specific implementation rules accordingly. The Department of Natural Resources of Zhejiang Province issued a notice, in January 2023, which specifically stipulated the organization, demonstration, review, report, daily supervision and other work of limited human activity construction projects involving land use approval and planning permission within the ERLs, in order to further strengthen the demonstration and report of construction projects with limited human activities within the ERLs.
2. Carrying out daily monitoring of ERL. Different technical means were used for the daily monitoring of ERL on land, marine and uninhabited islands, and monitoring results were analyzed and evaluated. For the land ERLs, the annual change survey data were used in monitoring the land use structure in the ERLs of the province (Fig. 5.8). The remote sensing change detection technology was used to extract the map polygons within the annual ERLs. For example, forest lands were transformed into roads and rural roads, and cultivated lands, forest lands and water areas were transformed into filled earth for construction. It carried out

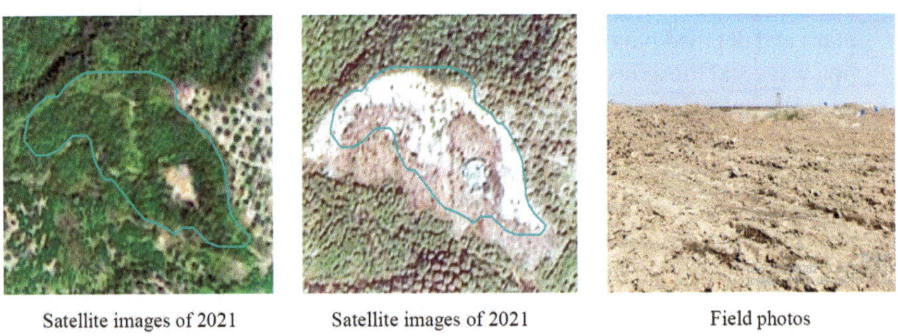

Fig. 5.8 Geomorphic change within the ERLs

an analysis of suspected new types of human activities, such as suspected newly increased filling earth, suspected newly increased construction (structure), suspected newly increased transportation land to provide technical support for timely detection of artificial activities within the ERLs. For the marine ERLs, human activity area, seawater environmental quality and sediment environmental quality within the ERLs were monitored and evaluated by means of remote sensing information extraction and field sampling measurement. For the uninhabited island ERLs, the conservation rate of important biological species in typical uninhabited islands, sea level rise and other conditions were monitored by using field investigation and other methods to effectively grasp the overall ecological environment quality and ecological conservation status of ERL in the province.

3. Improving the intelligent supervision level of ERL through digital reform. With the guidance of digital reform, based on the Zhejiang provincial spatial governance digital platform, the Zhejiang provincial territorial and spatial planning-ERL implementation monitoring and early warning platform has been developed. The platform includes digital cockpit platform, visual large screen, monitoring backstage management system and other module functions, achieving "ERL control of important ecological space" through digital means. The platform integrated the map patches of ERL through information integration, intelligent early warning of human activities, visual expression of approval results, unified coordination of law enforcement personnel in provinces and counties, and closed-loop management of illegal behaviors, so as to promote timely monitoring, early warning and law enforcement capacity of the natural resources department for key ecological function areas, sensitive and vulnerable areas of the ecological environment, improved the daily supervision of ERL, and enhanced the ability to respond quickly to emergencies.

5.5 Anhui Province: Consolidating the Foundation of the Green Chang Jiang Delta[6]

The ERL framework is a vital institutional innovation for ecological conservation in China. Anhui Provincial Government has delineated and published the ERLs in 2018. To deal with the contradictions & conflicts caused by human activities within the ERL area, Anhui Provincial Government has made an evaluation and adjustment of ERL, which has explored and improved the control mechanisms for ERL,

[6]This part was written by Pei Fulong, Wang Dong, Liu Bangrui and Chen Zhirui. Pei Fulong, the clerk of the Land and Space Planning Department, the Department of Natural Resources of Anhui Province; Wang Dong, the vice president and senior engineer of Anhui Territorial and Space Planning Institute; Liu Bangrui, an engineer of Anhui Territorial and Space Planning Institute; Chen Zhirui, an engineer of Anhui Provincial Natural Resources Information Center.

according to the unified national deployment, considered social economic and development needs.

5.5.1 *Overview of Delineation*

Anhui Province is located in the inner part of the Chang Jiang Delta, along the middle and lower reaches of the Chang Jiang and Huai He. It is in the Huang He ecological zone and the Chang Jiang & the Sichuan-Yunnan ecological zone. The entire province is divided into five natural areas including plains, hills and mountainous areas. According to the importance of ecological functions such as water conservation, biodiversity conservation, soil erosion control etc., Anhui Provincial Government has delineated 19,600 km^2 ERLs through evaluation and adjustment, which are mainly distributed in the Dabie Shan in western Anhui, mountainous areas in southern Anhui and plains along the Chang Jiang. The ecological spatial feature “one heart, two barriers, four corridors and multiple nodes” (“one heart” is the ecological green area of the Chao Hu; “two barriers” are ecological security barriers in western and southern Anhui; “four corridors” are the ecological corridors of the Chang Jiang mainstream and the wetland along the Chang Jiang, the Huai He mainstream and the wetlands along the Xin’an He, water diversion project from the Chang Jiang to the Huai He, and the Xin’an He mainstream and the areas along the Chang Jiang; “multi-node” is a variety of protected areas) has been constructed, which further consolidates the land ecological division of China’s six crucial ecological zones.

5.5.2 *Delineation Practices*

5.5.2.1 Overall Arrangement

Anhui Provincial Government has attached great importance to the evaluation and adjustment of the ERLs, and set up a technical guidance group to conduct “Dual Evaluation”, formulate relevant technical standards, and provide technical guidance to the localities to conduct evaluation and adjustment of ERL. Taking the results of “Dual Evaluation” as support, the local counties (cities) have conducted internal analysis and field reconnaissance, according to the requirements of evaluation and adjustments which brought the integrated and optimized protected areas within the scope of the ERL framework, thus forming the results of evaluation and adjustment at the county (city) level. The results of ERL evaluation and adjustment have been formed at the provincial level after expert argumentation (Fig. 5.9).

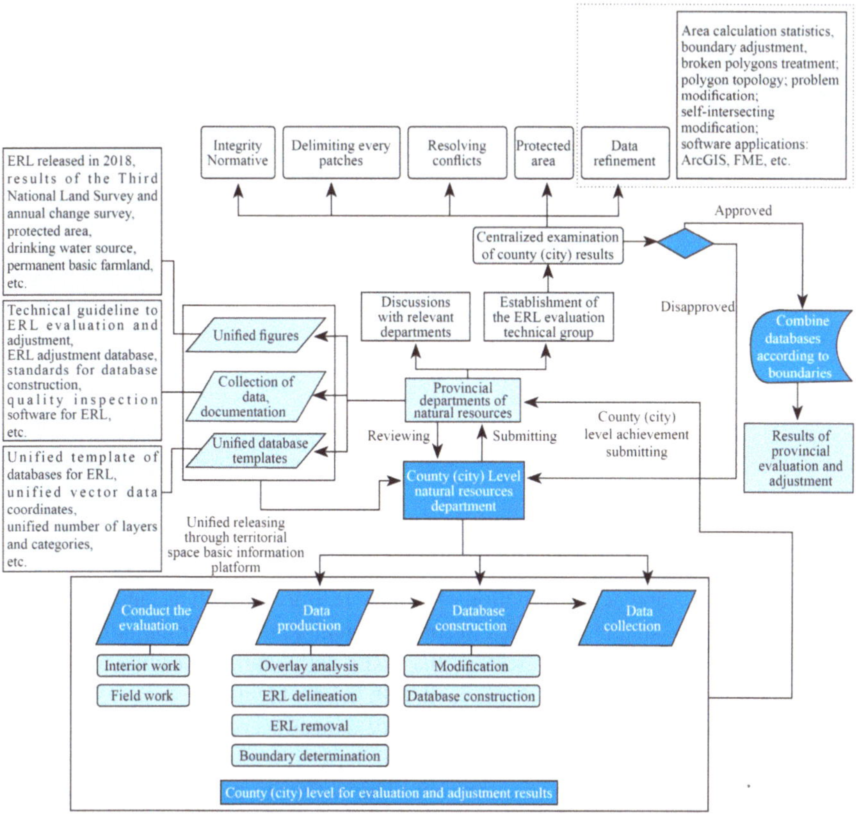

Fig. 5.9 Technology flowchart

5.5.2.2 Delineation Principles

1. Adhering to the priority of environmental conservation. Guided by maintaining ecological functions and ensuring territorial ecological security, the provincial government has implemented the main functional zones, conducted the provincial "Dual Evaluation", docked with national governance results, and verified the delineation of the ERLs. It has brought the vital ecological zones, as supplementary plots, within the scope of the ERLs framework, built up China's ecological security, and ensured that all eligible areas were delineated into the ERLs, highlighting the key regions and fields, focusing on areas with contradictions & conflicts in the delineating process.
2. Persisting in seeking truth from facts. Anhui Provincial Government has conducted scientific evaluation, after coordinated consideration of the integrity and systematicness of natural ecology, and made full use of the results of the Third National Land Survey and its annual change survey in Anhui Province (Fig. 5.10), so as to form a unified working base and base maps. Based on the distribution

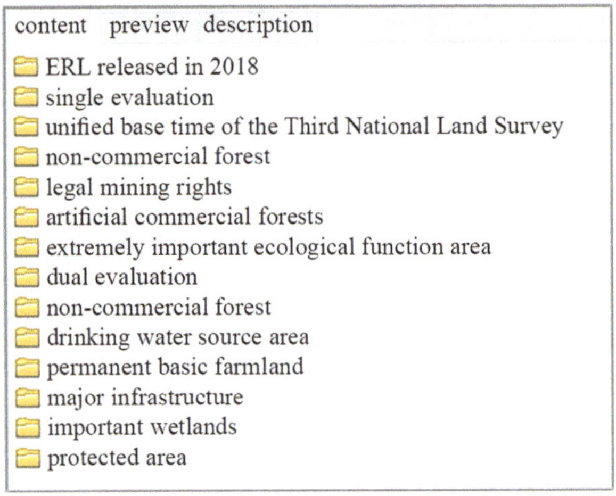

Fig. 5.10 Basic data

characteristics and actual situation of natural resources, such as landscapes, forests, lakes, grasslands and sands in Anhui Province, the government has made a review and analysis of the results of the ERLs delineation according to the rules and policies. It has classified and dealt with the problems left over from history, and focused on the problems of multi-source data convergence and conflicts arising from human activities.

3. Adhering to coordinated advancement. The authority at provincial level have defined and adjusted the rules. Provincial, municipal and county governments have carried out evaluation and adjustment of the ERLs, according to the technical specifications and policy requirements, through close coordination and communication. On the one hand, it has closely coordinated the integration and optimization of protected areas, focused on coordinating and solving the conflicts arising from the permanent basic farmland and mining rights in protected areas, and incorporated the integrated and optimized protected areas into the scope of the ERLs framework. Meanwhile, it has coordinated the designation of permanent basic farmland, dealt with the contradictions & conflicts between ERLs within the scope of urban construction, and reserved space for the delineated boundaries for urban development.
4. Adhering to overall stability. On the basis of the results of the delineated ERLs, and the premise of ensuring that "the ecological functions remain", Anhui Provincial Government has made the required adjustments and improvements to the ERLs according to the evaluation and adjustment rules, and proposed the adjustment and supplementary schemes to keep the general pattern at a stable level.

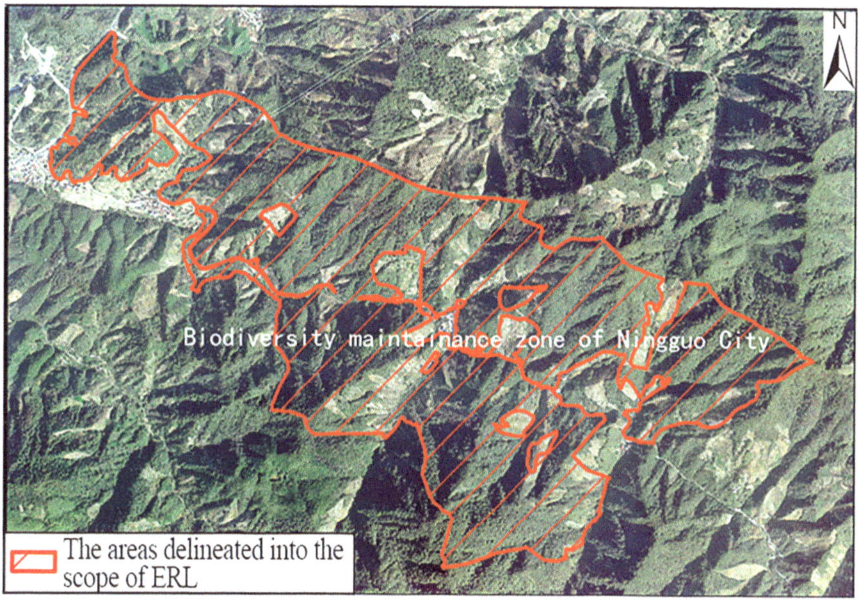

Fig. 5.11 The area of extremely important ecological functions within the scope of ERL

5.5.2.3 Refinement of Rules

Anhui Provincial Government amended the rules for evaluation and adjustment, which are set out as follows:

1. Checking and filling in gaps. The vital zones for ecological conservation determined by "Dual Evaluation", as well as the first-class conservation areas of the newly added and adjusted centralized drinking water sources at or above the county level, will be given priority in the scope of the ERLs. The first-class national non-commercial forests and natural forests that were not brought within the scope of the ERL framework due to incomplete basic data should be added, as well as the areas such as Chang Jiang, Huai He, Xin'an He, Chao Hu. Anhui Provincial Government has coordinated the integration and optimization of protected areas, and brought the integrated and optimized protected areas within the scope of the ERLs framework (Figs. 5.11, 5.12, 5.13, and 5.14). The areas within the scope of ERL covered an area of more than 4000 km^2.
2. Resolving conflicts. According to the relevant national regulations, the major infrastructure used for human activities should remain within the scope of ERLs, in order to ensure the systematicness, integrity and connectivity of the ERLs. For the areas outside the scope of the ERLs, any contradictions or conflicts except the permitted human activities should be accurately identified, and resolved by adjustment of the ERLs or withdrawal from human activities.

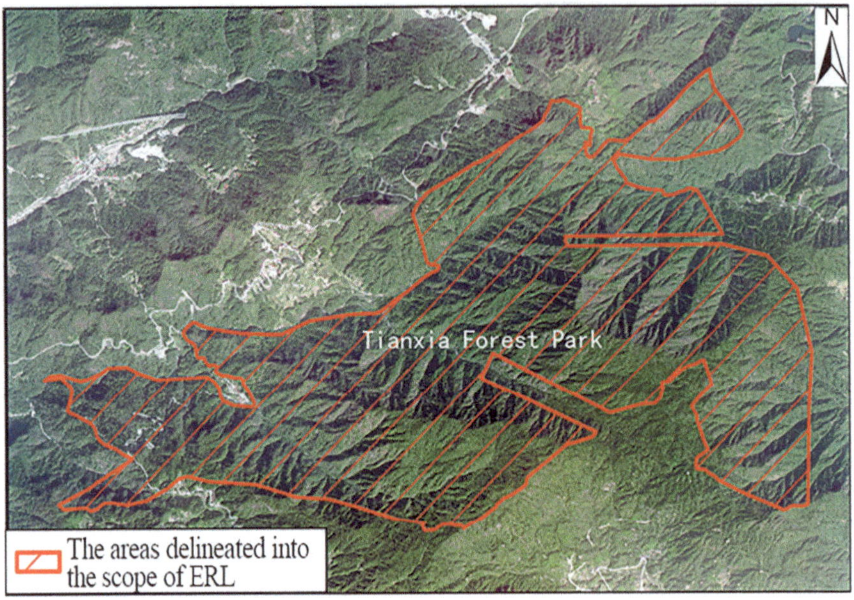

Fig. 5.12 The area of protection within the scope of ERL

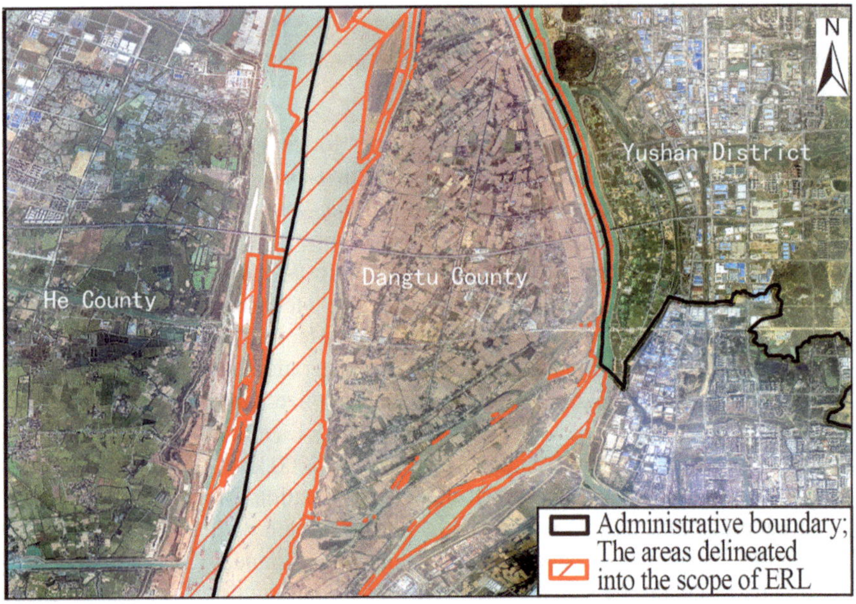

Fig. 5.13 The Chang Jiang within the scope of ERL

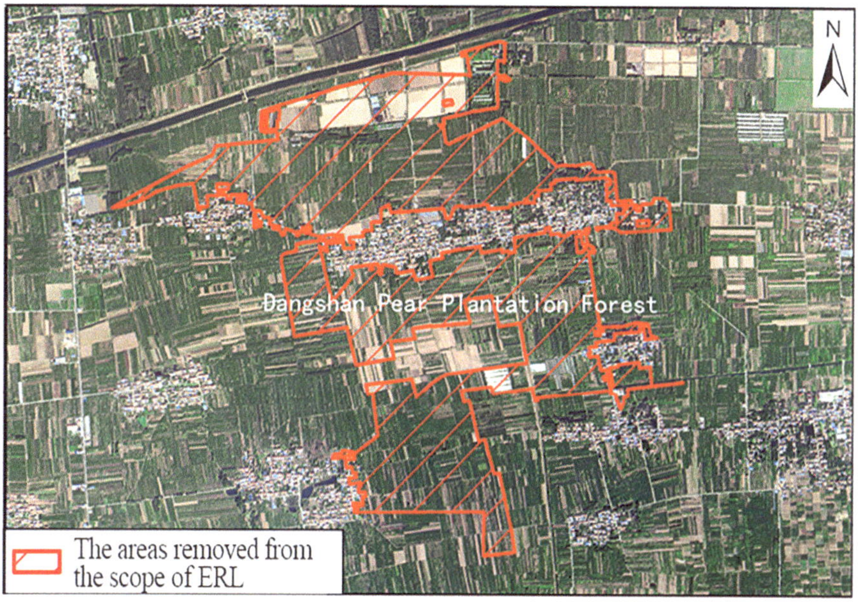

Fig. 5.14 The Dangshan Crisp Pear economic forests removed from the scope of ERL

3. Spatial coordination. Following the principles that "Three Zones & Three Lines" of permanent basic farmland, ERLs, and boundaries for urban development should not cross each other, overlap or conflict, the government departments should ensure spatial coordination, give priority to cultivated land and permanent basic farmland, identify the conflicts between the ERLs and the designated permanent basic farmland, and make timely adjustments. The overlaps between the protected areas for centralized drinking water source, permanent basic farmland, and urban construction, at or above the county level, shall not be included in the scope of ERL.

5.5.2.4 Delineation Process

1. Activating coordination. During the evaluation and adjustment of the ERLs, the government conducted relevant studies and formulated the rules for ERL management. To comply with the laws, regulations and rules, Anhui Provincial Government has actively coordinated with related departments for instructions, overcome difficulties, reported the results of ERL evaluation and adjustment at each stage, and made continuous improvement.
2. Training organization. Anhui Provincial Government has convened provincial mobilization and deployment and technical training workshops, unified thoughts and working ideas, conveyed the provincial work program, and held meetings

for the ERLs evaluation and adjustment to solve the existing problems in local work.

3. Base maps unification. Anhui Provincial Government has collected data from related departments, such as the results of the Third National Land Survey and the late annual change survey, and the data of protected areas, rivers and lakes, wetlands, forests and major projects etc. It organized the data conversion for the inconsistent technical standards, formulated unified work base maps, made data overlaying, concerning designated ERLs, protected areas for centralized drinking water sources, at or above the county level, legal mining rights, development zones and other related vector data, and published the data to local cities.
4. Elaborate implementation. Anhui Provincial Government has convened work promotion meetings and technical review meetings, promoted the evaluation and adjustment of work in an orderly manner, and kept informed of the work progress and difficulties in local cities. The government has carried out evaluation and adjustment according to the division of tasks and the work requirements, organized personnel to conduct on-the-spot reconnaissance, sorted out the contradictions & conflicts, analyzed the reasons, and proposed solutions according to the requirements.
5. Platform support. Relying on the provincial data technology groups, Anhui Provincial Government has provided special technical support from data reception, quality inspection, library integration, etc., and checked the results of delineation within the province.
6. Centralized review. To ensure the quality of the results reported by localities, members of the leading group have been organized to conduct centralized review of the evaluation and adjustment results reported by local cities, to review the key zone maps one by one, and to resolve the problems (if any) without delay. For those which cannot be resolved on the spot, the government put forward opinions, and submitted the problems to the localities for settlement within a time limit. Based on the integration and optimization results of the protected areas, Anhui Provincial Government has made dynamic adjustments, thus ensuring that the integrated and optimized protected areas have been brought within the scope of the ERL framework.

5.5.3 *Major Achievements*

5.5.3.1 Consolidating Ecological Spatial Patterns

The area within the ERLs in Anhui Province accounts for 14% of its territory which includes key ecological sources, linear corridors and ecological nodes, and covers the important ecological spaces within the province. The result leads to a more concentrated and optimized ecological spatial pattern, well implementing the national strategy for ecological conservation pattern. The spatial pattern of the ERLs has been further optimized, with Dangshan Pear economic forests in north Anhui removed, and Chang Jiang, Huai He, Xin'an He and Chao Hu were delineated into the ERLs. The ecological spatial pattern of "one heart, two barriers, four corridors and multiple nodes" was featured.

5.5.3.2 Improving Ecological Services

The ERLs of Anhui Province include all integrated and optimized protected areas. The areas with important ecological functions, such as water conservation, soil erosion control, flood diversion and storage, including the Chang Jiang, Huai He, Xin'an He, Chao Hu, mountainous areas in southern and western Anhui, Jianghuai hilly areas, plains in northern Anhui, as well as those areas with key biodiversity conservation functions, were brought within the ERL framework. The optimized ecological spatial pattern can effectively maintain Auhui's ecological functions, such as biodiversity conservation, water conservation and soil erosion control, thus leading to overall enhancing the ecosystem services.

5.5.3.3 Improving Human Settlement Environment

Higher regulatory requirements on ecosystem service guarantee, ecological security and natural resource utilization etc., were raised so as to promote the balance of population, resources and environment and the unity of economic, social and ecological benefits. Through evaluation and adjustment, Anhui Provincial Government made the ERLs more realistic, which effectively connected environmental pollution control and environmental quality improvement with environmental risk prevention, so as to ensure that ecology quality was maintained and gradually improved, and to assist in building a beautiful home with blue sky, green land and clean water, and providing more ecological products to meet people's growing needs for a beautiful ecology.

5.5.3.4 Promoting Sustainable Economic and Social Development

To coordinate the relationship between ecological conservation and the development of the economy and society, Anhui Provincial Government has carried out evaluations and adjustments to deal with contradictions & conflicts within the scope of ERL because of the historical problems and realistic development on a scientific and objective basis. It has coordinated the needs in reality and future, balanced the relationship between ecological conservation and development, so as to promote intensive and effective spaces for working, livable spaces for living and a green ecology without prejudice to ecosystem services. It guided economic layout to adapt to the carrying capacity of resources and the environment, thereby enhancing the ecological carrying capacity for sustainable economic and social development.

5.5.4 *Promoting the Implementation Practice*

5.5.4.1 Strengthening the Conservation and Control of Major Rivers and Lakes

Anhui Province, with a focus on ERLs, protected areas and important habitats for aquatic organisms in the Chang Jiang Basin, strictly conserved rare and endangered wild animals and biodiversity in the Chang Jiang Basin. Anhui Provincial Government strengthened water pollution control in Huai He, ensured that new construction, reconstruction and expansion projects complied with the relevant provisions of water pollution prevention regulations and were controlled in the Huai He Basin, and strengthened the aquatic genetic resources conservation in the Huai He Basin. The provincial government strictly conserved the natural resources, human resources and ecology of the Xin'an He and its banks, and maintained the landscape characteristics of southern Anhui. The government implemented three-level conservation for the water environment in Chao Hu Basin, strengthened the water pollution prevention and control in Chao Hu Basin, maintained the ecological services of Chao Hu Basin, strictly conserved the ERLs.

5.5.4.2 Accelerating the Construction of Huangshan (Guniujiang) National Geological Park

Giving full play to the advantages of cultural and natural resources in southern Anhui, the government focused on strengthening the conservation of natural ecosystems, natural relics and natural landscapes, and steadily promoted the construction of Huangshan (Guniujiang) National Geological Park. Anhui Provincial Government has implemented the national requirements, strengthened the management and control of the areas where plant and animal species are distributed, with the core value of Huangshan (Guniujiang) National Geological Park as the candidate, and improved the conservation of the cross-regional protected areas like Guniujiang and Qingliangfeng national protected areas, as well as the ecological spaces, such as key water sources and forest parks. It protected key spaces, such as national parks and surrounding ERLs as a whole.

5.5.4.3 Maintaining the Ecological Conservation Pattern in the Chang Jiang Delta Region

Anhui Province strengthened the construction of green ecological barriers in the Dabie Shan in western Anhui, the barriers in southern Anhui-western Zhejiang-southern Zhejiang and the coordinated control of transboundary rivers, such as Xin'an He-Qiandao Hu and Hongze Hu. Anhui Provincial Government worked with the neighboring provinces such as Zhejiang and Jiangsu to delineate the ERLs of the region and boundaries for river conservation, and continued to promote the pilot work of ecological compensation in the Xin'an He Basin. Anhui Provincial

Government provided ecological compensation in the upper and lower reaches of the Chu He Basin and Tuo He Basin, signed the cooperation framework agreement on ecological conservation of Shijiu Hu with Jiangsu Province, and focused on consolidating the green development of the Chang Jiang Delta and the delineation of demonstration zones for building a Beautiful China.

5.5.4.4 Strengthening the Overall Ecological Conservation and System Restoration

Anhui Province adhered to the concept of respecting, adapting and conserving nature, followed the principles of giving priority to economizing, conservation and restoration, comprehensively considered various factors in natural ecosystems, including top or foot of mountains, land ground or underland, upper or lower reaches and both banks of rivers, and implemented integrated conservation and restoration programs. Anhui Provincial Government strengthened the ecological restoration of the forest and grassland, improved the quality and stability of the forest ecosystem. It promoted the ecological restoration of rivers and lakes, strengthened the restoration and management of shoreline resources, implemented mine geological environment management and restored the ecology of mining areas, thus carrying out comprehensive treatment of coal mining subsidence area.

5.5.5 Strengthening Management in the Future

5.5.5.1 Building a Supervision System in Which the Governments Coordinate with Each Other

Governments at all levels, as the main body responsible for adhering to the ERLs, should take it as key criteria and prerequisite for relevant integrated decision-making processes, fulfill their responsibilities for ecological conservation, strengthen the construction of coordination mechanisms, form the coordination mechanisms for supervision and management of human activities within the scope of the ERLs, with the participation of related departments. The government should solve major problems in the work, guide the supervision and management of human activities within the scope of the ERLs, strengthen dynamic monitoring by means of information technology and information sharing among departments, and observe the rules of ERL together.

5.5.5.2 Improving the Control Over Access by Classification and Zoning

Anhui Provincial Government should carry out zoning control within the scope of the ERLs, and further refine the scope permitted for human activities, except for major national strategic projects, on the premise of compliance with current laws and regulations. A control system to prohibit human activities within the scope of the ERL should be set up, which focuses on reviewing the compliance of project types, the unavoidability of site and route selection, economic and intensive use of land. For the prohibited human activities which have existed, a removal mechanism should be established to guide the removal step by step.

5.5.5.3 Implementing Differentiated Policies of Ecological Restoration

Ecological restoration in the core reserves should focus on natural restoration, supplemented by necessary artificial measures. It is important to build ecological corridors, carry out the restoration of key habitat and abandoned land, and strengthen manual intervention for ecological damage caused by mining and invasion of alien species. For other regions, the government will arrange the required ecological restoration, along with the improvement of ecological services and the orderly withdrawal of human activities. The government should strengthen the joint prevention and control of ecological pollution in the ERLs and surrounding areas, and focus on the comprehensive governance of rivers and lakes within the scope of the ERLs.

5.5.5.4 Establishing the Monitoring-Based Mechanism for Periodic Evaluation and Adjustment

Based on the results of monitoring and evaluation of territorial and spatial planning, the government should carry out evaluation and control of the ERLs regularly. When the ecosystem has changed greatly due to major strategic adjustment, project implementation and ecological changes, and an adjustment is required, the authority should formulate a plan for adjustment and submit it to the State Council for approval according to procedures before implementation. To ensure the management and control of the ERLs, it is prohibited to adjust by modifying the city, county and township plans without authorization. For qualified provincial capitals or cities designated by the State Council, the plans for adjustment, together with the revision of plans, can be submitted to the State Council for approval before implementation.

5.5.5.5 Promoting an Atmosphere for Joint Participation in Ecological Conservation

Anhui Provincial Government will intensify policy propaganda, carry out publicity activities on particular days, such as “World Environment Day” and various media platforms, create a good social atmosphere, further enhance people’s awareness of ecological conservation. Anhui Province will ensure timely and accurate release of the information concerning the ERLs’ distribution, adjustment and conservation according to the requirements, and safeguard the public’s right to know, participate in, and supervise ecological conservation.

5.6 Guangdong Province: Guarding the Pearl of Lingnan[7]

Guangdong Province has a tropical and sub-tropical monsoon climate. This region, rich in light, heat and water resources, has both land and marine. Its landscape is diverse, with high terrain in the north and low terrain in the south. The northern Guangdong, which is a mountainous area composed mainly of the Nanling Mountains, is a key zone for biodiversity conservation in China, a key ecological barrier in the Guangdong—Hong Kong—Macao Greater Bay Area (GBA) and a key component of territorial ecological security. The southern Guangdong, with a vast sea area, has a long coastline, nice bays, gulfs, tidal flats, and numerous islands. The coastal zone performs the functions of coastal conservation and marine biodiversity conservation. Three rivers converge in the region, forming a developed river system, with a panoramic view of mountains, rivers, forests, farmlands, lakes, grasslands and sands. The province's natural ecosystems cover lands, wetlands and marine, with a wide variety of animals and plants and biodiversity characteristics. The delineation of the ERLs in a scientific manner plays an important role in exploring economic construction and ecological civilization construction, coordinating development paths, thus building a new pattern of harmonious development between man and nature, and a world-class bay area.

5.6.1 Delineation Practices

1. Process of delineation. According to the national work deployment, Guangdong Provincial Government carried out the work of the ERL delineation. The plan was reviewed by the provincial government executive congress in August 2018, and reported to the State Council for approval. The work program of the ERLs delineation for sea areas was approved and implemented by the provincial government in September 2017.

   In June 2019, MNR and MEE carried out evaluation and adjustment of the ERLs, and issued a series of documents, which defined the procedures, standards and requirements of the audit work.

   In June 2021, the program of the ERLs evaluation and adjustment of Guangdong Province was submitted to the State Council. In July 2021, Guangdong Province, as one of the five pilot projects designated by the national

[7] This part was written by Tang Hui, Li Junting, Luo Hongming, Liu Cuixia and Chen Ling. Tang Hui, the deputy director of the Land and Space Planning Department, the Department of Natural Resources of Guangdong Province; Li Junting, the director of the Land and Space Planning Department, the Department of Natural Resources of Guangdong Province; Luo Hongming, the director of the Comprehensive Office of Guangdong Provincial Land Survey and Planning Institute; Liu Cuixia, an engineer of Guangdong Provincial Land Survey and Planning Institute; Chen Ling, the deputy director of the Planning Office, Guangdong Provincial Land Survey and Planning Institute.

"Three Zones and Three Lines", explored the adjustment of the ERLs. In April 2022, MNR issued the document on delineation of permanent basic farmland boundaries, ERLs, and urban development boundaries. The results of delineation were put into use with the approval of the State in October 2022, which served as the basis for the examination and approval of development on land or sea.

2. Principles of delineation. In the process of delineating the ERLs, Guangdong Provincial Government has always adhered to the following principles:

First, bottom-line thinking on ecological development, taking ecology as a priority. To implement the national requirements, Guangdong Provincial Government practiced the concept of ecological priority, maintained ecological functions, ensured ecological security, conserved the places as much possible as can, and improved the integrity, systematicness and connectivity of natural ecology, thus building the bottom line of ecological conservation.

Second, evaluation in a scientific manner, delineation as much as possible. On the basis of the vital important protected areas through "Dual Evaluation" for territorial space, based on the natural boundaries, such as mountains, rivers, landforms and vegetation, the scope of the ERLs has been identified in a scientific manner, thus ensuring rational layout and clear boundary of the ERLs.

Third, seeking truth from facts, keeping the delineated ERLs. Guangdong Provincial Government adhered to the problem orientation, followed the principles that "Three Zones & Three Lines" in permanent basic farmland, the ERLs, and boundaries for urban development do not cross each other or overlap, by focusing on solving the problems left over from history and realistic contradictions & conflicts within the scope of ERL, thus ensuring the reasonable development space of the cities in the future. In addition, the ERLs can be implemented and kept firmly.

Fourth, arranging multi-party coordination and linkage from top to bottom. To make an overall plan to promote the evaluation and adjustment of the ERLs and the integration and optimization of protected areas, Guangdong Provincial Government brought the results of integration and optimization of protected areas into the scope of the ERLs in a timely manner, complied with the national requirements, and strengthened communication and coordination with related provincial government departments, extensively solicited opinions and suggestions from local governments of counties, thus reaching the maximum consensus.

Technology flowchart of delineation in shown in Fig. 5.15.

Guangdong Provincial Government brought the key ecosystems and protected areas within the scope of the ERLs framework, such as vital important protected areas through "Dual Evaluation", the ecological conservation pattern, national and provincial prohibited-development areas, mangroves, etc. It kept ecological security barrier and built the ecological security bottom line for future generations.

First, unifying the base map and laying a solid foundation. Guangdong Provincial Government has adopted the results of the Third National Land Survey, the results of the Second-class Survey of Forest Resources, and the

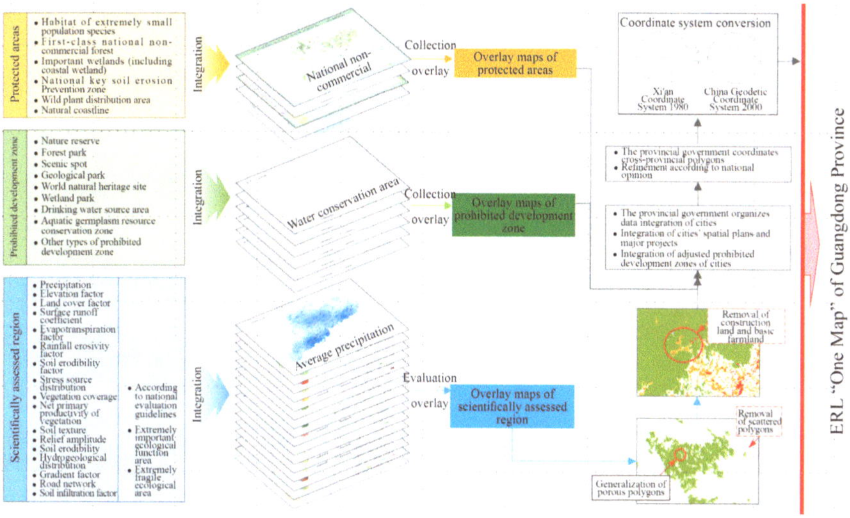

Fig. 5.15 Technology flowchart for ERL delineation of Guangdong Province

results of Coastline Updating Survey according to the national requirements, so as to ensure the authenticity and accuracy of basic data and the compliance with national data results.

Second, doing addition in a scientific way, delineating the ERLs as much as possible. Guangdong Provincial Government, as the first batch of "Dual Evaluation" demonstration provinces in China, insisted on seeking truth from facts, studied and selected the models and algorithms that best reflect the reality of Guangdong, and carried out "Dual Evaluation" of important water conservation, biodiversity conservation, soil erosion control and sensitive grading evaluations. It evaluated and identified the vital important ecosystem service zones covering an area of more than 1 km^2 with no contradiction or conflict, summarized the zones providing key or potentially key ecosystem services, which need to be strictly protected, and brought these zones within the scope of the ERLs (Fig. 5.16). It dynamically coordinated the integration and optimization results of protected areas, brought the results of integration and optimization of protected areas within the scope of the ERLs, so as to ensure that key spaces for ecological conservation are delineated as much as possible.

Third, subtracting on an objective basis and dealing with contradictions & conflicts. To avoid the risk that may be difficult to manage when delineated exactly according to the polygon, Guangdong Provincial Government has carried out "Dual Evaluation" according to the requirements of the national technical guidelines and the ecological conservation pattern issued by the state, and brought the added vital ecosystem service zones within the scope of the ERLs framework, and treated the remaining areas causing conflicts or plots covering an area of 1 km^2 or less, and ensured that such areas are not brought within the

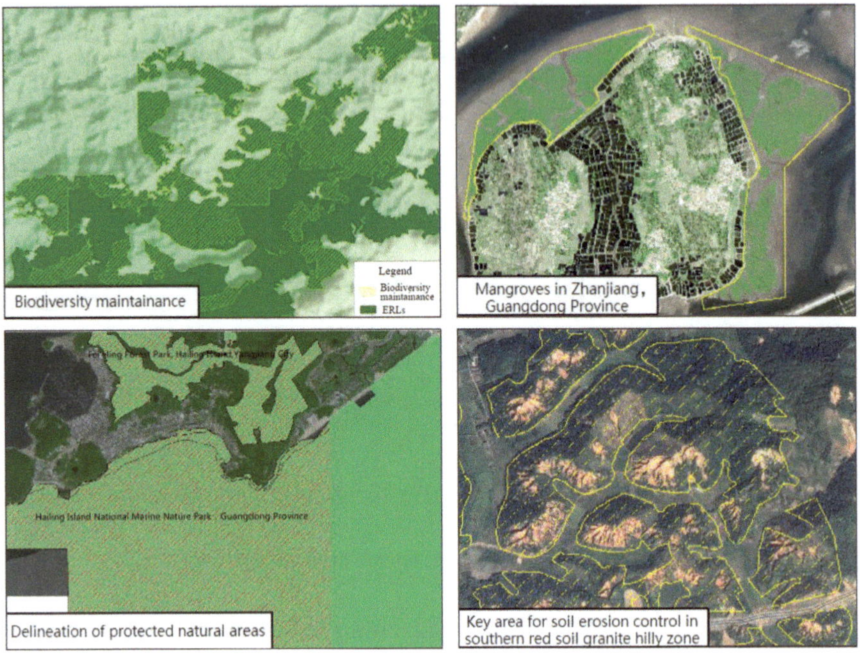

Fig. 5.16 The areas within the scope of ERL of Guangdong Province

scope of the ERLs framework. Moreover, during the delineation, the territorial conflicts arising from the permanent basic farmland and legal mining rights, towns and villages, artificial commercial forests and construction projects have been systematically sorted out and verified. The related government departments have convened three rounds of consultation and coordination, and formulated working rules for evaluation and adjustment of the ERLs, which defined the rules for coordinating and handling conflicts. More than 10,000 conflicts reported by local cities and departments have been handled, which supported local departments to solve problems related to legal mining rights, land-reclamation, artificial commercial forests, as well as historical and realistic contradictions & conflicts, through which the local cities were left a reasonable space for future development (Fig. 5.17).

Fourth, making overall plans for development and security, and strengthening the work link-up. (1) Ensuring the delineation and link-up with the "Three Zones & Three Lines". Confirming the priority order of cultivated land, permanent basic farmland, the ERLs and boundaries for urban development, Guangdong Provincial Government made sure that the three lines of control do not cross each other, overlap or conflict. Adhering to the priority of cultivated land conservation, on the premise of ensuring no obvious impact on ecosystem services, Guangdong Provincial Government had the contiguous spot of no less than 3 mu (approximately equal to 2000 km^2) cultivated land that was stably used for a long

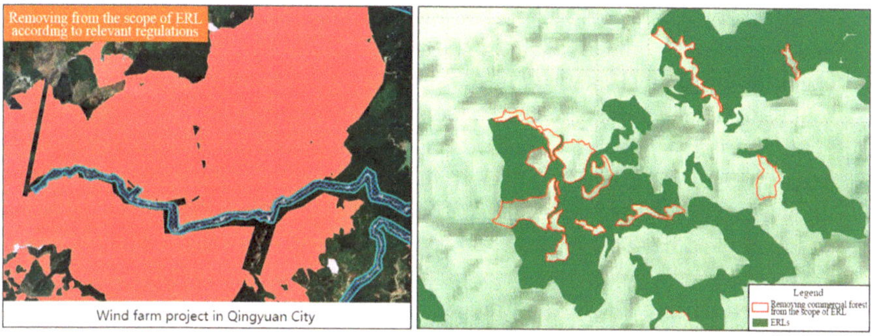

Fig. 5.17 Dealing with conflicts of ecological conservation in Guangdong Province

Fig. 5.18 Dealing with conflicts between ecological conservation and permanent basic farmland in Guangdong Province

time, and outside the core reserves of the ERLs removed out, and converted it into permanent basic farmland (Fig. 5.18) to ensure that permanent basic farm land was delineated as much as possible. The land that is being used or planned to be used for cities, towns and villages outside the protected area, and the planned areas for mining, major projects and infrastructure, etc., should be removed from the ERLs or not included in the ERLs, thus reserving the construction space. (2) Promoting the overall planning of land and sea, ensuring the link-up with Coastline Updating Survey. The ERLs in Guangdong Province have adopted the latest Coastline Survey results as the boundary to define the land area, and the overlapping areas of land and marine are brought within the scope of sea. The sea areas with typical marine ecosystems, such as mangroves and coral reefs, as well as unauthorized, unused and uninhabited islands in coastal waters, islands where territorial sea base points were located and islands within the conservation scope of territorial sea base points, uninhabited islands for national defense purposes, uninhabited islands in marine protected areas, uninhabited islands with special conservation values, such as rare and endangered wild animals and plants, habitats and important natural relics, etc., would be brought within the scope of the ERLs. The integrity and natural connectivity of coastal ecosystems were fully considered to realize the effective connection of land and sea ERLs.

Fifth, combining field work and internal work, strengthening field surveys and providing proof. Guangdong Provincial Government carried out internal check and technical analysis in the process of delineation, as well as several rounds of field reconnaissance to understand the authenticity of contradictions & conflicts on the spot, and discussed and exchanged ideas with local government departments and the masses to determine the principles and suggestions for dealing with contradictions & conflicts. The cities like Zhuhai, Yangjiang, Shaoguan and Qingyuan have been selected as typical coastal cities and mountainous cities, respectively. With the help of high-resolution remote sensing images, unmanned aerial vehicles and related technical means, Guangdong Provincial Government made an in-depth study of typical problems, aiming to provide more accurate data and scientific basis for the delineation of ERLs.

5.6.2 Spatial Pattern of the ERLs

The ERLs in Guangdong Province distribute as a pattern of "two ecological security barriers and one shelterbelt".

1. The ERLs of land area in Guangdong Province mainly include five types, namely, water conservation, soil erosion control, biodiversity conservation, soil and water loss control and rocky desertification control. Those were specifically divided into 14 zones, including five water conservation zones, three soil erosion control zones, three biodiversity conservation zones, one soil and water loss control zone, and two rocky desertification control zones.

   The ERLs of land area included key areas of water conservation and biodiversity conservation, such as the Fenghuang-Lianhua Mountains in the east, Nan Ling in northern Guangdong and the Yunwu Mountains in the west, and the extremely fragile areas of soil erosion and rocky desertification in the Nan Ling in north, and Pingtianzhang-Lianhua Mountains in the east of Zhu Jiang Delta, Huadu-Conghua Mountains in the northern Zhu Jiang Delta and Tianlu Shan in the western Zhu Jiang Delta, forming a "two-barriers" ecological conservation pattern consisting of a circular ecological barrier in northern land area and an ecological barrier around the Zhu Jiang Delta.
2. The ERLs of marine area in Guangdong Province were divided into 12 categories, including mangroves, coral reefs, seagrass meadows, important estuaries, important tidal flats and shallow water areas, specially conserved islands, distribution areas of rare and endangered species, spawning grounds of important fishery resources, extremely important areas for coastal biological conservation, extremely important areas for coastal physical conservation, ecologically fragile areas for coastal erosion and ecologically fragile areas for sand loss. The ERLs of marine area were mainly distributed in Nan'ao, Nanpeng Islands and Chaoshan coastal wetlands in eastern Guangdong, Dongshan, Zhu Jiang Estuary, Daya Bay, Shenzhen Bay and Qi'ao-Dangan Island in the GBA. Hailing Bay, Shuidong

Bay, Tongminghai, Xuwen and Leizhou in western Guangdong, and coastal zones with good contiguous beaches and mangrove coasts in western Guangdong, forming the pattern of three ecological zones and one shelter belt.

5.6.3 Management Thoughts in the Future

5.6.3.1 Implementation in Territorial and Spatial Planning

According to the national deployment requirements, Guangdong Provincial Government has insisted on overall conservation and development, implemented the strict system of ecological conservation, determined agricultural space, ecological space and urban space in the overall planning of territorial space in a scientific manner, prohibited adjusting the ERLs, conserved ecological space, maintained the delineation results of "Three Zones & Three Lines", and ensured that no adjustment should be made to the delineation results of the ERLs in urban and county planning of territorial space. It has regarded the area of the ERLs as a binding target of territorial and spatial planning, which was transmitted and implemented layer by layer in the preparation of territorial and spatial planning at provincial, city, county and township levels and incorporated into the "one map" information supervision system of territorial and spatial planning. Guangdong Provincial Government has promoted the consistency of the ERLs in base map, data and database, realized the information management and control of it, strengthened the awareness of ecological conservation, and taken the ERLs as the rigid constraint of project site selection, so as to protect the ERLs in a rational way. Guangdong Provincial Government insisted on forbidding all construction activities that destroy the ecology and violate the requirements of ERL control.

5.6.3.2 Strictly Controlling the Examination and Approval of Land and Marine Use

Guangdong Provincial Government strictly implemented the relevant requirements of MNR and MEE, standardized and controlled limited human activities within the scope of the ERLs, controlled the territorial space occupation of major national projects, and handled the examination and approval of land, sea and island use, in strict accordance with regulations. The development and productive construction activities were prohibited outside the core reserves within the scope of the ERLs. On the premise of compliance with the current laws and regulations, only limited human activities were allowed by the national legislation. Guangdong Provincial Government was requested to issue opinions on the identification of limited human activities in strict accordance with procedures. According to the current situation of the province, the government has formulated the notice on managing the ERLs after study, and further refined the workflow by allowing limited human activities and major national projects involving new construction on land and sea islands to occupy territorial space within the scope of the ERLs, strictly defined the adjustment, and strengthened the supervision and management.

5.6.3.3 Exploring the Management and Control of the ERLs

1. Further improving the relevant policies and mechanisms. According to the national management measures of the ERLs, the government will study and formulate the regulatory measures for the ERLs in Guangdong Province, further improve relevant policies and mechanisms, clarify local territorial management responsibilities and requirements, and promote governments at all levels to abide by the ERLs.
2. Establishing a dynamic adjustment mechanism for the ERLs. It is supposed to establish a collaborative linkage mechanism for the integration and optimization of protected areas. If protected areas are adjusted according to laws and regulations, the authority of natural resources management should make synchronous adjustments to the ERLs according to the approval documents, update the "one map" of territorial and spatial planning, ensure data coordination, and promote the integration and optimization of scattered eco-land to improve the integrity and connectivity of the ecosystem. The authority should add the "skylight" spots which human activities have withdrawn to the scope of the ERLs after the regular review and examination of territorial and spatial planning so as to enhance the integrity and connectivity of the ecosystem.
3. Strengthening the long-term mechanism of the ERLs supervision. According to the requirements of unifying base map, standard, planning and platform, the results of the ERLs delineation were collected and integrated into the "one map" of territorial and spatial planning step by step, and information sharing was realized with the national ERL ecological supervision platform, which served as the key basis for territorial and spatial planning implementation supervision and ecological monitoring. Guangdong Provincial Government strengthened the real-time sharing of data and results among organizations, and enhanced the modernization of spatial governance. It will improve the monitoring and supervision, and accelerate the construction of efficient, full-coverage, full-process and all-factor ERL monitoring and conservation evaluation system. According to the requirements of regular examination and "5-year evaluation", it is necessary to establish and improve the monitoring, evaluation and early-warning mechanisms for the implementation of territorial and spatial planning at all levels, and carry out an evaluation of the results of ERL regularly, so as to refine and improve the ERL rules based on the evaluation results.
4. Guiding the public to participate in supervision. It needs to publicly release the report on the status of the ERLs in a timely manner, encourage the public and social organizations to participate in and supervise the management and improve the attention of all sectors of society to the ERLs, so as to incorporate the delineation scope and related information into the "Map World" system and open to the public. The public may inquire whether the planned urban development,

mining, construction project layout, planning and layout, road route selection, etc., involve the ERLs anytime and anywhere, so as to effectively protect the ERLs at the source.

5.7 Sichuan Province: Ecological Barrier in the Upper Reaches of the Chang Jiang and Huang He[8]

The ERLs are the bottom line and lifeline of safeguarding and maintaining territorial ecological security. Sichuan Provincial Government further strengthened the awareness of ecology and shouldered its responsibilities, according to the national unified deployment on delineating and enforcing the ERLs in a coordinated manner. Sichuan Provincial Government has refined the rules of management and control, as well as conducted supervision and management, thus realizing the management of all major environmental spaces with consistent standards, and had the spatial foundation of key ecological barriers in the upper reaches of the Chang Jiang and the Huang He consolidated.

5.7.1 General Ideas for the ERLs Delineation

The authority unified the base map based on the Plans for Ecological Red Lines of Sichuan Province issued by Sichuan Provincial Government in 2018, carried out an in-depth evaluation of ecosystem service importance and ecological vulnerability, coordinated the results of integration and optimization of protected areas, focused on identifying existing contradictions & conflicts, clarified adjustment rules, and carried out targeted evaluation and adjustment.

5.7.1.1 Delineation Principles

1. Giving priority to conservation and highlighting key goals. Guided by maintaining ecological functions and ensuring the ecological security of the state and the province, the government implemented the strategy and system of functional zones, focused on evaluating and solving the contradictions & conflicts from the delineation of the ERLs, and identified the areas that should be delineated, so as to ensure the ERLs delineation as much as possible.

---

[8] This part was written by Zheng Yuan, Gou Bin, Feng Han, Xia Taiyun and Gao Huanggen. Zheng Yuan, the director of the Regional Research Institute of Sichuan Institute of Territorial and Spatial Planning; Gou Bin, the deputy director of the Regional Research Institute of Sichuan Institute of Territorial and Spatial Planning; Feng Han, an engineer of the Regional Research Institute of Sichuan Province; Xia Taiyun, the vice president of Sichuan Space Planning Institute; Gao Huanggen, the president of the Sichuan Institute of Territorial and Spatial Planning.

2. Seeking truth from facts, being scientific and objective. By establishing a sense of the bottom line, complying with working procedures, quality requirements and auditing standards, the government made overall consideration of the integrity and systematicness of nature ecology, carried out scientific evaluation, and found out the background of ecological resources in the province, so as to carry out the delineation of ERL in a scientific and objective manner.
3. Unifying the base and realizing coordination. Making full use of all types of objective survey data, based on the recognized results of the Third National Land Survey, the provincial government ensured the preparation of territorial and spatial planning and the integration and optimization of protected areas, thus realizing the overall connection, regional coordination and cooperation between departments.
4. Keeping the overall level stable and achieving local optimization. On the premise of ensuring that ecosystem services are not reduced, the area of ecosystem service zones has not decreased, and the nature remained the same, on the basis of scientific evaluation, the ERLs were partially adjusted and optimized, according to the idea of "addition first, subtraction later".

5.7.1.2 Main Practices

1. **Consolidating the foundation and ascertaining the actual situation**
   In the field analysis, making full use of the basic data such as high-resolution remote sensing images, the data from the Third National Land Survey, and the integrated and optimized protected areas, Sichuan Provincial Government analyzed and evaluated the accuracy of basic data, the rationality of delineated boundaries, contradictions & conflicts, and delineations as much as possible, so as to ascertain the actual situation of ERL in the province. Meanwhile, adjustment rules were updated according to the local conditions. The rules of adjustment and control of the ERLs in Sichuan Province have been formulated.
2. **Coordination between departments, optimization and perfection**
   On the basis of the process which is up five times and down five times, the Department of Natural Resources of Sichuan Province has coordinated with the Forestry and Grassland Bureau of Sichuan Province, Giant Panda National Park, Chinese Academy of Sciences and the Institute of Mountain Hazards and Environment to promote the integration and optimization results of protected areas and the optimization of "Dual Evaluation" results. On the basis of delineation as much as possible in the ERLs, it coordinated with the offices of in charge of development and reform, water resources, communications and transformation to resolve the contradictions & conflicts within the scope of the ERLs, so as to ensure that the delineation of ERL is scientific and feasible.
3. **Reviewing software to improve quality and efficiency**
   Through auxiliary software, the provincial government checked the data results of the ERLs submitted by counties patch by patch. It put forward problems regarding conflicts between the ERLs and other land use types, as well as

fragmentary patches in maps and irregular boundaries in vector data etc., which provided data support for scientific evaluation of the ERLs, reducing repeated workload in evaluation and adjustment, and improving the quality and efficiency of ERL evaluation and adjustment.

4. **Perfect linking and overall planning**

   Sichuan Provincial Government linked the results of integration and optimization of protected areas with the results of "Dual Evaluation" of ecological conservation to formulate the evaluation report of the ERLs. It was revised and improved according to the opinions of related departments of the province, and reported to the State. According to the preliminary review and technical review by the MNR, the revised results of the ERLs were filed by MNR to various Ministries and Commissions for countersigning, and were finally approved by the State Council. After the promulgation and implementation, the boundaries were delineated based on the territorial and spatial planning at city and township levels. The technology flowchart is shown in Fig. 5.19.

5.7.2 Optimizing the Ecological Conservation Pattern

Sichuan Province is located in the upper reaches of the Chang Jiang, in the inland hinterland of southwest China, which is a transition belt from the first step of staircase to the second step of the staircase in China's topography. The alpine and plateau region in the west and the mountainous region in the southwest are key components of the ecological conservation pattern of "two barriers and three shelterbelts"—Qinghai-Xizang Plateau Ecological Barrier and Loess Plateau-Sichuan-Yunnan Ecological Barrier respectively. The main ecological functions are water conservation, soil erosion control, biodiversity conservation and climate regulation. The region serves as the ecological barrier in the upper reaches of the Chang Jiang and the Huang He, and the key components of "water tower".

According to the relevant requirements, the ecologically sensitive areas in the mountains and plateaus of western Sichuan, the sensitive areas of soil erosion in the lower reaches of the Jinsha Jiang and the sensitive areas of rocky desertification in southeastern Sichuan, and other ecologically important or fragile areas have been brought within the scope of the ERLs. Other ecological spaces that are not brought within the scope of the ERLs are conserved by a system that suits their conditions. Through these measures, the ecological space both inside and outside of the ERLs is conserved incorporately.

The ERLs in the province are mainly distributed in the areas such as mountains, plateaus and the basin-mountain, in the northwest of Sichuan, and reflected in the spatial layout of "four axes and nine cores". The "four axes" include Daba Shan, dry-hot river valley in the lower reaches of the Jinsha Jiang, mountainous areas and hilly areas in the basin in the southwest, and the ERLs are mainly distributed in belts. The "nine cores" include Zoige Wetland (Huang He Source), Yalong Jiang Source, Dadu He Source, Daxue Shan, Shaluli Shan, Min Shan, Qionglai Shan,

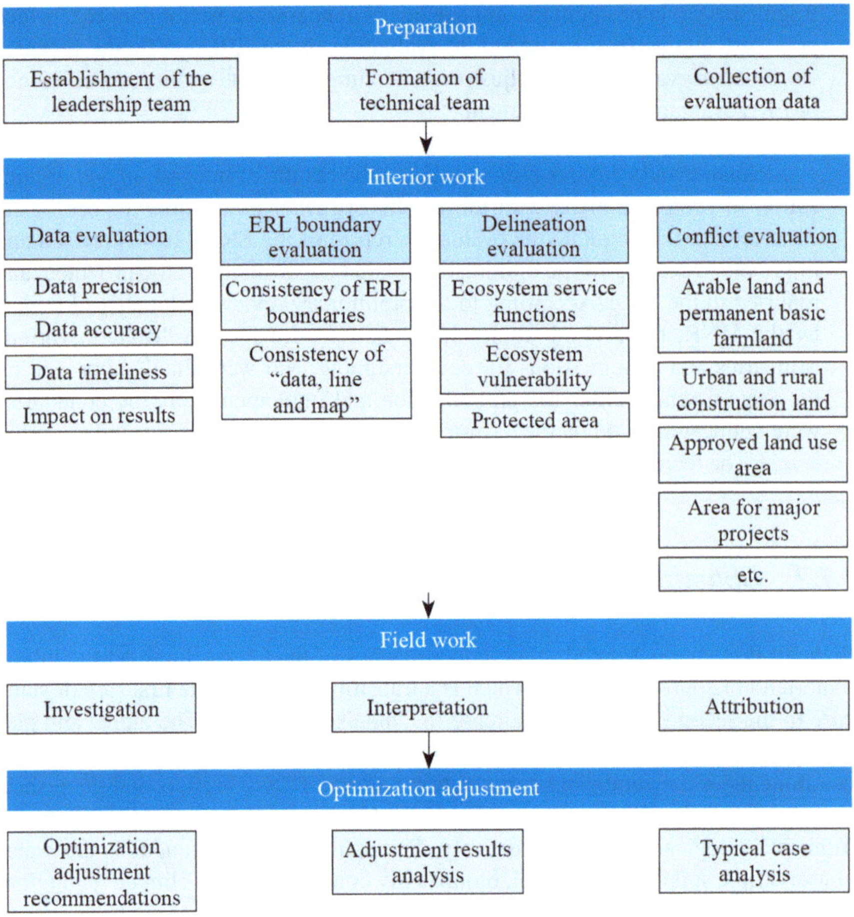

Fig. 5.19 Technology flowchart of ERL evaluation for Sichuan Province

Liang Shan-Xiang Ling Mountain and Jinping Shan, and the ERL feature concentrated distribution with water system and mountain system as the skeletons.

5.7.3 Strengthening the Conservation and Restoration of Important Ecological Systems

5.7.3.1 Establishing a Nature Protected Area System with National Parks as the Main Body

Sichuan Provincial Government accelerated the establishment of a protected area system featuring national parks as the main body, nature reserves as the foundation and nature parks as the supplement, and continued to improve the quality of the ecosystem to ensure overall ecological security.

Giant Panda National Park has been built with high quality. Taking the panda conservation as the core ecological function, keeping the authenticity and integrity of the natural ecosystem as the basis, and adhering to the concept of respecting, adapting to and conserving nature, the delineation of the ERLs provided a good ecological space for the survival and reproduction of pandas and their umbrella-protected animal and plant species to ensure the stable reproduction of the panda population.

Zoige National Wetland Park had been built. It was co-established by Sichuan and Gansu provinces. The core goals of the establishment are to conserve the authenticity and integrity of the plateau peat swamp wetland eco-system, protect the ecology of important habitats of rare birds such as black-necked cranes, improve the ecosystem quality and water conservation capacity of the upper reaches of the Huang He, and consolidate the largest "reservoir" in the upper reaches of the Huang He.

Nature reserves construction and management have been strengthened. The core goals are to maintain and restore the population of rare and endangered wild animals and plants and the habitat environment on which they live, ensure the safety of the main protected objects.

The ecological conservation of various nature parks has been improved. The government strengthened the conservation of important natural ecosystems, natural relics and natural landscapes, as well as the zones for sustainable utilization with ecological, ornamental, cultural and scientific values, improved the ecological conservation functions of nature parks, and promoted the combination of conservation and utilization.

5.7.3.2 Building a Biodiversity Conservation System

Adhering to the priority of conservation and natural restoration, following the laws of natural ecosystem succession and zonal distribution, Sichuan Provincial Government strengthened the construction of the biodiversity conservation system, and continued to optimize the biodiversity conservation spatial pattern. It conserved key areas for biodiversity, and strengthened the conservation of priority areas for biodiversity conservation, such as the southern section of Hengduan Mountains, the northern section of Min Jiang-Northern part of Hengduan Mountains, Qiangtang-Sanjiangyuan, Daba Shan and Wuling Shan.

Sichuan Provincial Government conserved rare wild animal ecological corridors and bird migration passages, strengthened the construction of key ecological corridors for pandas in an all-round way, speeded up the construction of wild animal migration foraging passages, and strengthened the conservation of important aquatic organism migration passages and bird migration passages.

Sichuan Provincial Government improved the layout of species ex-situ conservation system, strengthened the ex-situ conservation management of animals and plants, and optimized the construction of the zoo and botanical garden and wildlife rescue network.

Sichuan Provincial Government strengthened the layout of the gene preservation system, promoted the construction of forest germplasm resource bank and grass seed resource bank for the conservation of plant genetic resources diversity, and built and improved wild animal rescue, breeding institutions and germplasm gene banks for the conservation of wild animal genetic resources.

5.7.3.3 Integrated Conservation and Restoration

Taking key river basins and mountains as basic units, highlighting the characteristics of natural geographical integrity, ecosystem connectivity and similarity of ecological problems, the province area was divided into eight zones for ecological restoration. Each zone has particular key goals and measures by considering the priority of conservation and natural restoration. There are 10 major projects of integrated conservation and restoration which have been implemented. During the process, Sichuan Provincial Government coordinated to solve outstanding ecological problems in the region, restored damaged ecosystems, improved ecosystem quality, and enhanced carbon sequestration capacity, thus promoting biodiversity conservation. Strictly abiding by ERLs, Sichuan Provincial Government consolidated the ecological conservation pattern, and strived to guarantee ecological security of ecosystem service zones.

5.7.4 Paying Attention to the Detailed Implementation of the ERLs

5.7.4.1 Unifying Base Map and Base Data

There were problems with the basic data used in the delineation of the ERLs, such as diverse data sources, different accuracy, insufficient timeliness, different projection and coordinate systems, etc., which led to many contradictions in the delineation results and made the delineation unable to be carried out. In the process of territorial and spatial planning at city and township levels, it was clarified that former ERLs should be adjusted based on the result of the Third National Land Survey, as well as the basic data such as high-resolution remote sensing images, carrying capacity of resources and environment and suitability of territorial space development.

In the process of delineation of the former ERLs for Hongya County, the boundary of red lines was jagged due to the poor accuracy of data such as protected areas. The ERLs were planned to be revised in the scale of 1:10,000 or more precise, in consideration of factors such as geomorphic boundary and the integrated and optimized scope of protected areas, so as to ensure the scientific and accurate delineation results of the ERLs (Figs. 5.20 and 5.21).

Fig. 5.20 Serrated diagram of the former ERLs of Wawushan Forest Natural Park in Hongya County

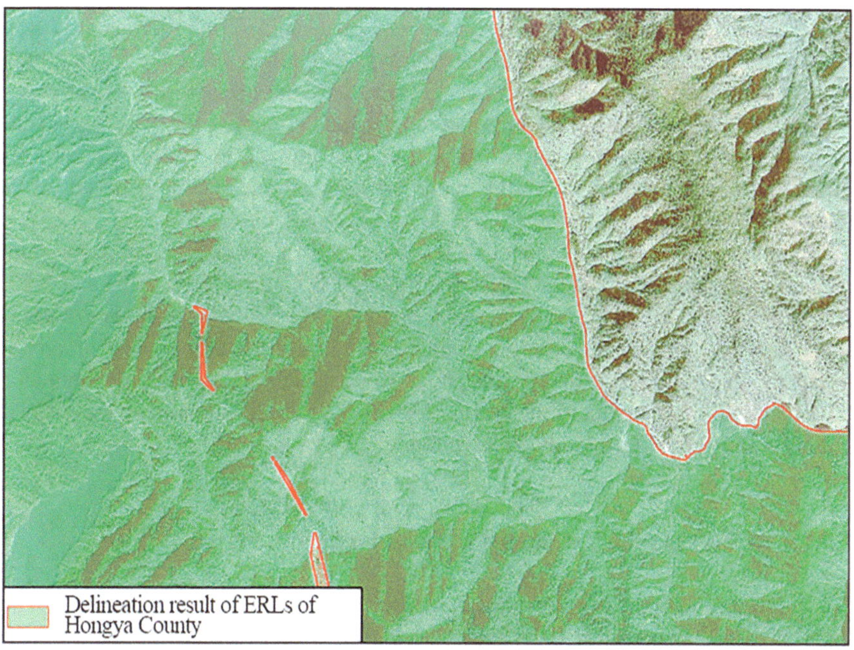

Fig. 5.21 Regional diagram of Wawushan Forest Natural Park, as delineation results of ERLs for Hongya County

5.7.4.2 Giving Priority to Ecological Conservation and Delineating into Red Lines as much as Possible

According to the carrying capacity of resources and the environment and the suitability evaluation of territory development, Sichuan Provincial Government coordinated the data such as the integrated and optimized protected areas, first-class conservation areas of drinking water sources, aquatic germplasm resources conservation zones, the integrity and systematicness of natural ecosystems and distribution areas of rare and endangered species etc., and brought the areas in need of strict ecological conservation within the scope of the ERL framework to ensure the integrity of the ecosystem and the connectivity of ecological corridors.

Baoxing County, based on the integrated and optimized scheme for protected areas, has brought the protected areas such as Giant Panda National Park, Baoxing Qiaoqi Hu Wetland Park, and Baoxing Rare-fish Nature Reserve within the scope of the ERL framework. The proportion of natural reserve delineated within the scope of ERLs has reached 100%. The proportion of ERL in the total area of the county has increased from 74.09% to 81.97%. The integrity of the ecosystem and connectivity of the ecological corridor have been significantly improved.

5.7.4.3 Resolving Conflicts and Contradictions

The government implemented the requirements of related documents of the State and Sichuan Province, analyzed and sorted out the contradictions & conflicts within the scope of the ERLs, such as permanent basic farmland, contiguous cultivated land, artificial commercial forests, current towns and villages, industrial and mining land, special land, approved construction land, and land used for key construction project, according to the actual situation of cities and counties. The government carried out targeted evaluation of ecosystem service impact, and adjusted and optimized the ERLs by means of adjustment, withdrawal and retaining based on the results of evaluation.

There is a case of adjustment. There is a patch of permanent basic farmland located in the scope of former ERLs for Jiuzhaigou County (Figs. 5.22 and 5.23). According to the result of the evaluation, Sichuan Provincial Government removed it because it is centralized, contiguous and outside the core reserve, with an area of 0.54 km^2. It coincides with the rule of "a slope of 25° below and an acreage of 0.01 km^2 above can be out of the scope of ERLs". After the adjustment, the piece of land is managed for farmland conservation.

5.7.4.4 Checking the Rationality of Boundaries

By comparing with the high-resolution remote sensing images and the results of the Third National Land Survey, the authority analyzed the rationality of the results of the ERLs delineation, rectified the ERLs with obvious deviations from geomorphic

Fig. 5.22 Diagram of the relationship between former ERLs for Jiuzhaigou County and centralized contiguous permanent basic farmland

units and administrative boundaries, and adopted cluster fusion of the rectified the ERLs via geographic information system, so as to reduce the "skylight" and the fragmentation of patch to ensure that the results of the ERLs delineation were consistent with natural geographical characteristics and ecological conservation pattern.

There was an inconsistency between the ERLs for Meishan City and the boundaries of physical geography before. Based on the boundaries of physical geography and the integrated and optimized scope of protected areas, the authority rectified ERLs to ensure the consistency of the ERLs map, actual quantity and physical object (Figs. 5.24 and 5.25).

Fig. 5.23 Diagram of the relationship between the delineation results of ERL for Jiuzhaigou County and centralized contiguous permanent basic farmland

5.7.5 Strict Control Measures for the ERLs

5.7.5.1 Current Status of Management and Control

1. **Controlling the limited human activities**

   First, strictly limiting the types of human activities causing no damage to ERL functions. The government specified that specific areas such as national parks, protected areas, natural parks, scenic spots, drinking water conservation areas, aquatic germplasm conservation areas, and World Heritage sites within the scope of the ERLs should be in strict accordance with laws, regulations and

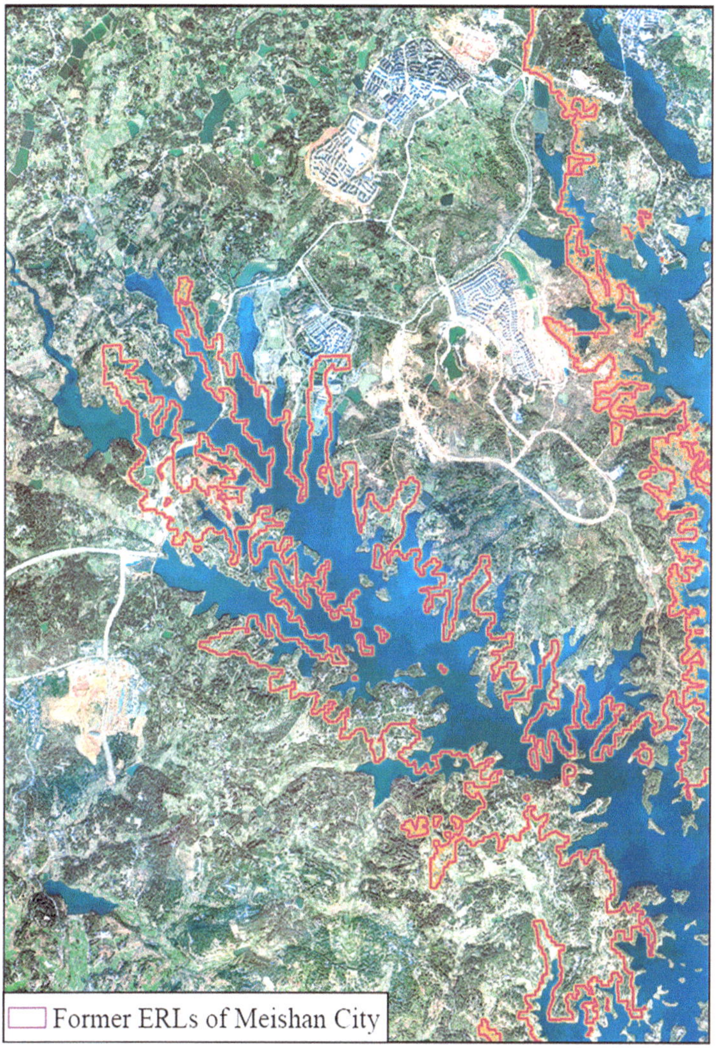

Fig. 5.24 Composite diagram of remote sensing images of former ERLs and Heilongtan Pool in Meishan City

provisions, and should get approval from the competent departments, before the land use is submitted for approval.

Second, standardizing the identification of limited human activities. If limited human activities with no damage to ecological functions need to be submitted for approval of land use, it is clear that the original land use approval would not change due to ERLs. With the consent of the provincial government, the process of identifying opinions on limited human activities within the scope of the ERLs was formulated as the basis for approval of agricultural land conversion and land expropriation.

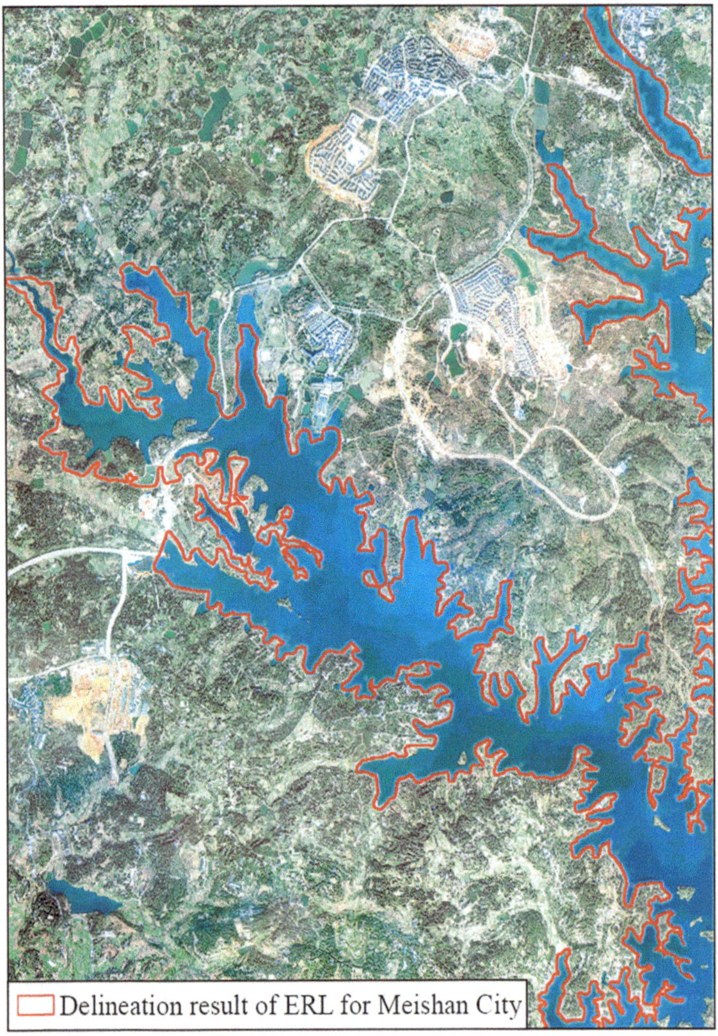

Fig. 5.25 Compositee diagram of remote sensing images of delineation results of ERL and Heilongtan Pool in Meishan City

2. **Strict examination and approval of land use/occupation**

   First, strictly controlling the scope of project. It clarified that the projects that occupy area within the ERLs must be limited to the scope stipulated in relevant documents, and projects that do not fall within the scope stipulated in official documents are not allowed to build and occupy ERLs.

   Second, standardizing the requirements for examination and approval. It stipulated the project basis and review opinions that should be provided when the land use is reported step by step, formulated the handling process of the opinions of demonstration on unavoidable occupation of the ERLs by major projects of the State, and clarified the content of the demonstration report on the unavoidable occupation of area within the ERLs by construction projects. The department of natural resources, the department of ecology and environment, and the forestry and grassland bureau, as well as relevant experts, reviewed and held arguments meeting before the land use is reported to the provincial government for comments.

Third, carrying out the work of "integrating multiple demonstrations". The government filed all documents, which included the arguments for site selection in the process of site selection for the proposed construction projects and pre-examination for land use, construction land reconnaissance and land-saving evaluation, and the demonstration of unavoidable occupation of the ERLs, on one-time, and synchronous approval to shorten the time limit and improve the efficiency.

3. **Strict supervision of the ERLs**

First, clarifying main responsibilities. It clarified the responsibilities of the related local departments, and required the departments at all levels to obey the ERLs together.

Second, dealing with the history remaining problems. It studied and drafted the document on withdrawal of mining rights from the scope of the ERLs, and promoted relevant work in an active and stable manner. It put forward principles and measures for disposal of artificial commercial forests, original residents and hydropower, wind power and photovoltaic power facilities scattered throughout the city.

Third, strengthening daily monitoring. The government strengthened the dynamic monitoring and evaluation of the ERLs, and notified the local government of the problems found in monitoring and control in a timely manner. Relevant departments verified on-site according to their respective functions, and dealt with them according to laws and regulations without delay. Cities and counties strengthened capacity building, information sharing, established a supervision system in their respective administrative areas, and implemented hierarchical supervision.

Fourth, insisting on strict procedures when adjusting. It put forward specific implementation opinions on the adjustment of the boundaries of protected areas and drinking water conservation areas, and the proposed transfer of oil and gas exploration rights established according to law.

5.7.5.2 Thoughts on Control and Management

Improving laws and regulations. Based on the current situation of Sichuan Province, relevant national laws, regulations, policies and measures, the government will formulate regulations according to local conditions. Sichuan Provincial Government will study and formulate plans and supporting policies for land use that are conducive to improving and safeguarding ecological services, encourage market-oriented and socialized investment and financing methods, mobilize resources from all parties as a whole, and explore multiple channels to raise funds, so as to jointly promote ecological conservation, restoration and governance.

Carrying out regular evaluations. The government will analyze and evaluate the ecological service status and dynamic change trend of the ERLs through regular evaluation, and take the evaluation results as the pre-link and important basis for optimizing the layout of ERL and arranging compensation funds for ecological conservation.

Establishing a monitoring platform. The government will incorporate the ERLs into the "one map" of territorial and spatial planning, integrate the monitoring stations such as ground ecosystem, environment, meteorology, hydrology and water resources, soil erosion control, and satellite remote sensing monitoring capabilities, improve the layout of the ERLs monitoring station network, and realize full coverage of the monitoring network. The government will improve the monitoring data sharing service system and mechanism, build a monitoring and early warning database, use cloud computing, big data processing and data fusion technology to realize real-time data sharing. The government will notify the local government of the destruction of ERL without delay, and the relevant departments will organize on-site verification according to their respective functions, and deal with the act of destroying according to laws and regulations.

5.8 Qinghai Province: Guarding the Chinese Water Tower[9]

The ERLs delineation of Qinghai Province is of great significance for stabilizing a territorial ecological security barrier, maintaining the national ecological conservation pattern, improving the ecosystem service of Qinghai-Xizang Plateau, and providing high-quality ecological products and promoting the green development and lifestyle.

5.8.1 Background

Qinghai Province is located in the northeast of Qinghai-Xizang Plateau, which is called the "roof of the world". It is the source of the Chang Jiang, Huang He, Lancang Jiang and other major rivers. Dubbed the "China's water tower" and "source of rivers", Qinghai is the hinterland of aquatic lifelines in the whole country

[9] This part was written by Tian Li, Kong Cuorui, Wang Yan, Liu Shuhui, Xu Xiang. Tian Li, the director of the Territoral and Spatial Planning Bureau, the Department of Natural Resources of Qinghai Province; Kong Cuorui, the director of the Development and Unilizaton Division, the Department of Natural Resources of Qinghai Province; Wang Yan, the chief of Qinghai Provincial Territorial and Spatial Planning Institute; Liu Shuhui, the vice chief of Qinghai Provincial Territorial and Spatial Planning Institute; Xu Xiang, an engineer of Qinghai Provincial Territorial and Spatial Planning Institute.

and even Asia and the key area of biodiversity of great global significance. As an ecological regulatory region with the greatest impact on the global atmosphere and water cycle, a key ecological barrier in the whole country and Southeast Asia, a start-up area and regulatory region of climate change in the Northern Hemisphere, a key wetland ecosystem in high altitude areas of the world, and biological germplasm resource gene bank for plateaus, Qinghai plays an irreplaceable and special role in maintaining the national ecological conservation pattern and building a territorial ecological security barrier.

A total of 296,500 km^2 have been delineated in the scope of the ERLs for Qinghai Province, accounting for 42.56% of the territory in Qinghai, covering integrated and optimized protected areas, vital important areas for ecological conservation and other areas that need strict conservation, effectively improved and conserved the connectivity, integrity and authenticity of natural ecosystems, optimized and improved the national ecological conservation pattern, and provided the insurance and maintained the bottom line of ecological security in the entire province even to the country. The ERLs presented a pattern of "one barrier, one shelterbelt and three ecological zones", with "one barrier" as the ecological barrier of Sanjiangyuan grassland, meadow and wetland, "one shelterbelt" as the Qilian Shan glacier and the ecological zone for water conservation, and "three ecological zones" as the ecological service zones for Qinghai Hu grassland and wetland, Qaidam desert and wetland and eastern hilly biodiversity respectively.

5.8.2 *Practices*

5.8.2.1 Working Ideas

On the basis of the Third National Land Survey, through carrying out "Dual Evaluation", the extremely important ecological function areas, extremely fragile ecological areas and the integrated and optimized protected areas, glaciers and permanent snow cover, drinking water conservation areas, etc., were identified to formulate the base map of the provincial ERL evaluation and adjustment. Based on the current situation of Qinghai and the need for future development, Qinghai Provincial Government has resolved the conflicts & contradictions among ecological conservation, economic and social development, according to the national rules, in a scientific, objective and factual manner, without prejudice to ecological functions.

5.8.2.2 Principles of Delineation

1. Conservation is given priority and the ERLs are delineated as much as possible. Guided by the implementation of territorial ecological security responsibility, Qinghai Provincial Government maintained the overall ecological conservation pattern of the province, gave priority to ensuring that extremely important eco-

logical function areas and extremely fragile ecological areas were delineated within the scope of the ERLs, and brought integrated and optimized protected areas at all levels within the scope of the ERLs framework.

2. Updating data and evaluating in a scientific manner. Taking the data of the Third National Land Survey as the work base and base map, based on the evaluation of the importance of ecological conservation and the integration and optimization of protected areas, the ERLs were delineated according to the actual situation to ensure the accuracy and scientificity of the delineation results.
3. Overall coordination and reasonable delineation. Qinghai Provincial Government followed the integrity and systematicness of the natural ecosystem, carried out scientific delineation, and coordinated the relationship between conservation and development based on the actual situation of Qinghai Province. It handled and resolved all historical problems and realistic contradictions & conflicts to ensure that the three controlled lines did not cross overlap or conflict with each other.
4. Coordination and linkage among departments. The Qinghai government strengthened the communication and coordination with local government and departments, thus enhancing the main responsibilities of local governments in setting and applying the ERLs. A mechanism interconnected upper and lower level governments has been set up which is beneficial to forming resultant force, so as to ensure that the ERLs can be delineated and maintained in a pragmatic way.

5.8.2.3 Technology Flowchart

The process is: First, to collect basic data of various industries and adopting a unified standard format, so as to make a base map of the ERLs evaluation and adjustment. Second, to analyze the boundary consistency, deal with the regions that need to be conserved. Third, to evaluate whether the delineation according to the principle of "delineated as much as possible". Fourth, to adjust the contradictions & conflicts within the scope of the ERLs according to the rules of evaluation and adjustment. Last, to improve the ERLs for Qinghai after listening to the opinions and suggestions of all parties, and forming the final result (Fig. 5.26). The experience of Qinghai is as follows:

1. Adjusting the collected data and unifying the base map and base data. Qinghai Provincial Government adopted the latest status data and planning data provided by the competent departments of various industries to update, and adjusted the source, type and accuracy. The data collection includes the integration and optimization plans for protected areas, ecological conservation importance evaluation, permanent basic farmlands, artificial commercial forests, transportation, water conservancy, mining rights, remote sensing images, etc. All data adopted China Geodetic Coordinate System 2000, which can improve the current situation and accuracy of the data, and refine the land use classification and ensure the authenticity, reliability and practicability of data source.

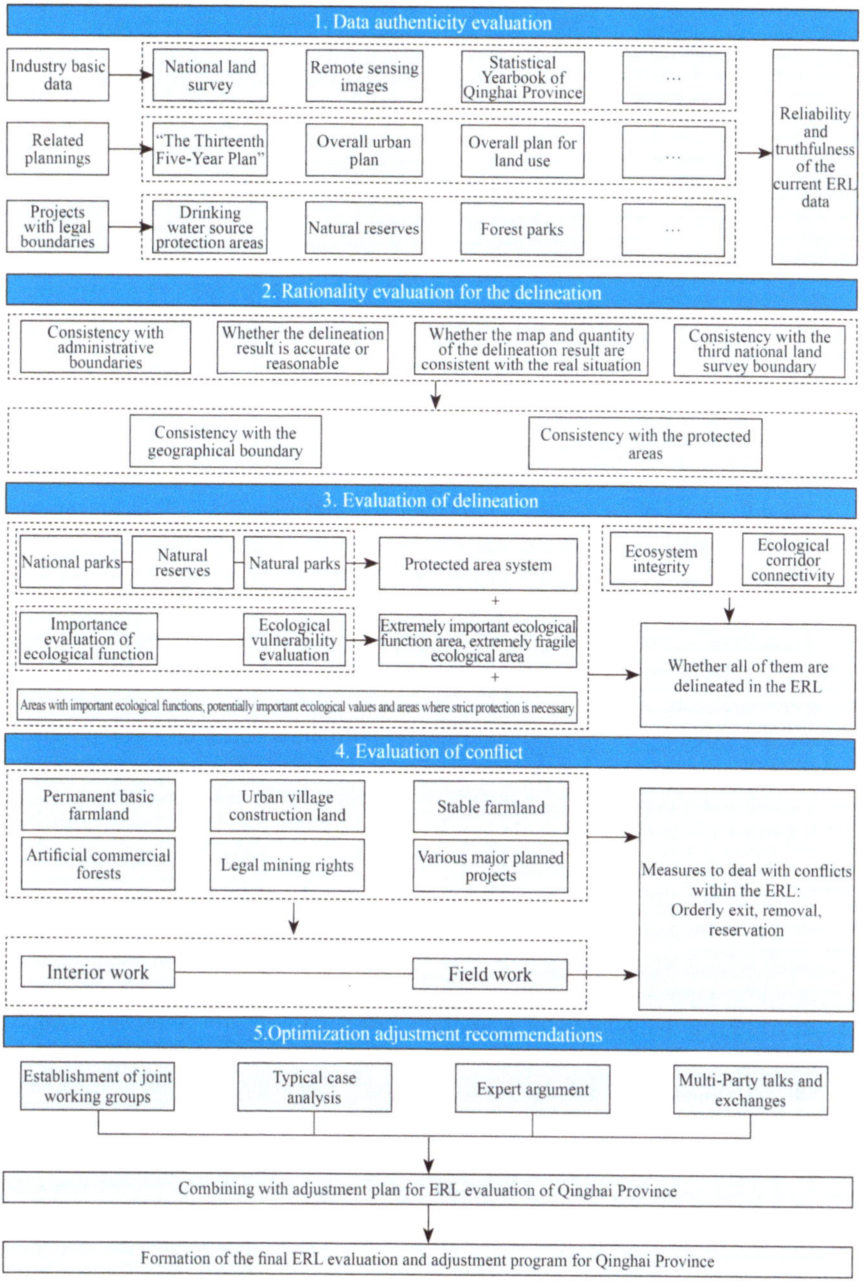

Fig. 5.26 Flowchart of evaluation and adjustment of ERL for Qinghai Province

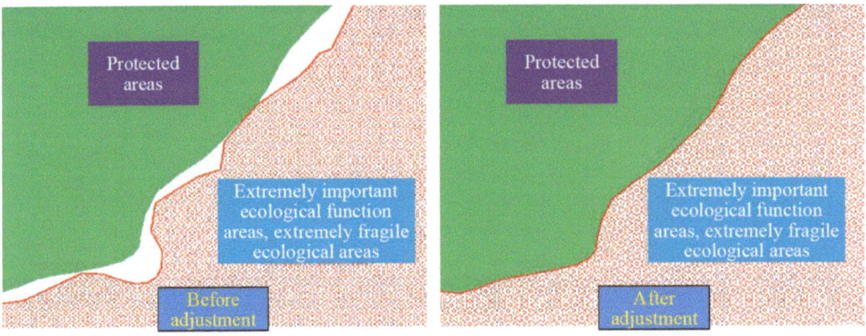

Fig. 5.27 Optimization diagram of boundary matching between protected areas, extremely important ecological function areas and extremely fragile ecological areas

2. Identifying key spaces and delimiting the ERLs as much as possible. Through "Dual Evaluation", it identified the vital important areas for ecological conservation of the province, and ensured that the evaluation results are fed back to the integration and optimization process of protected areas, which promoted the optimization of boundaries of protected areas, according to the results of "Dual Evaluation". 156,400 km^2 of vital important areas for ecological conservation were included in the protected areas and all the integrated and optimized protected areas were set in the ERLs. The area of protected areas within the ERLs has increased by 28,900 km^2, and the proportion has been up from 76.16% to 100.00%.

   100% of glaciers and permanent snow covers, 100% of international wetlands (Qinghai Hu Bird Island wetland nature reserve, Zhaling Hu wetland, Eling Hu wetland), 93.43% of first-class national non-commercial forests and 98.60% of Qinghai Hoh Xil World Natural Heritage were delineated into the ERLs, thus natural ecological spaces including forests, grasslands, wetlands and glaciers could be conserved as a whole, and the quality and stability of the ecosystem could be improved.
3. Revising the boundaries in a uniform manner to ensure rationality. Qinghai Provincial Government made boundary consistency revision of the ERLs, physical geographical boundary, protected areas boundary, administrative district boundary and provincial boundary, thus keeping them consistency in space. In addition, it delineated the ERLs as much as possible (Fig. 5.27).
4. Clarifying the rules for adjustment and institutionalizing the delineation. According to the clear adjustment rules of the state, the local government studied and issued relevant documents such as work plan, adjustment plan, technical guide. By adopting the model of interaction among the adjustment of protected areas, Qinghai Provincial Government clarified the adjustment rules for conflicts between major projects, farmlands, mining rights, towns and villages, residential areas, planned land use, etc. (Fig. 5.28), so as to maintain enough space for local economic development.

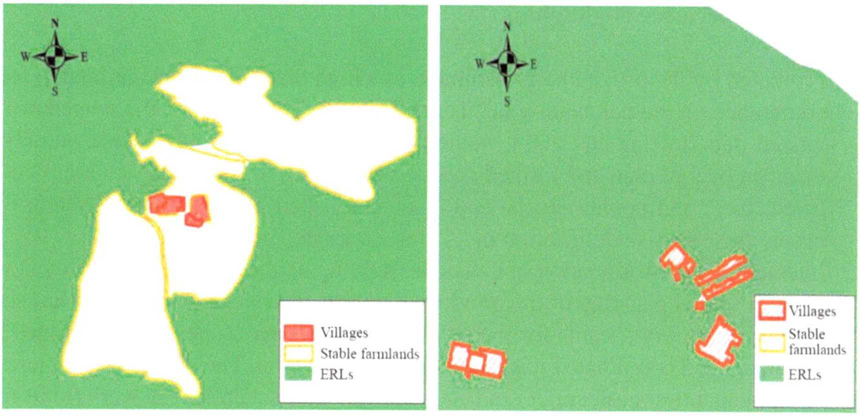

Fig. 5.28 Classification diagram for conflict resolution within the ERLs

5.8.2.4 Organizing the Implementation

1. Attaching great importance, making overall planning. The provincial, municipal and county governments have set up special work teams for the ERLs and established a joint meeting system. Governments at all levels and their relevant departments have adopted the method of "systematic planning, classified guidance and overall coordination", and made concerted efforts from top to bottom to get this work done.
2. Close cooperation and coordinated implementation. The provincial department of natural resources, department of ecology and environment and forestry and grassland bureau have cooperated closely and coordinated with each other. A series of normative documents such as relevant work plans, adjustment rules and technical guidelines were issued, which clarified the objectives, tasks and key points as well as rules and technical methods of the ERLs delineation for Qinghai Province, thus providing a guiding basis for the ERL delineation.
3. Active communication and effective docking. In order to give an overall consideration to the balance between ecological conservation and social and economic development in various regions and resolve the problems left over and the realistic conflicts on the basis of fact, the work team found out the actual condition of the ERLs, the rationality of boundary delineation, the delineation proportion and conflicts, etc., which laid a solid foundation for scientific delineation of the ERLs.

   Last but not least, the experts of MNR went to Qinghai several times, providing guidance on the delineation of ERLs, which supported this work powerfully.

5.8.2.5 Experience Highlights

1. Promoted by the provincial government at a high level. After the national video conference on the delineation of "Three Zones & Three Lines", the related provincial departments and eight municipalities (prefectures) have acted quickly, made an overall plan and formed special work teams at provincial, municipality (prefecture) and county levels, and also established a work contact system to ensure the effective delineation of "Three Zones & Three Lines".
2. Interaction and coordination of divisions. Relying on the leading group of territorial and spatial planning of the province, Qinghai Provincial Government has set up provincial work teams for the delineation of "Three Zones & Three Lines", organized and coordinated related affairs of delineation. The related departments of municipalities (prefectures) and counties have also set up special work teams to form joint efforts. The special work teams of the government departments at provincial, municipality (prefecture) and county level have established effective communication channels, coordinated internal resources, dispatched 1500 cadre employees focusing on tackling key problems, continued to play the guiding role of technical working teams, worked effectively together to promote the delineation.
3. Solicit opinions from all parties. The leader group of territorial and spatial planning of the province has conducted field visits and investigation in municipalities (prefectures) and counties, made a written request to related local departments to provide relevant information, and given feedback five times to related local departments on the conflicts between special plans and key projects, so as to deal with major problems and discrepancies. Meanwhile, the group properly resolved the contradictions & conflicts in the delineation process.
4. Joint review of the results. The government departments at the provincial level, together with the government departments at municipality (prefecture) and county levels have conducted several rounds of county reviews, organized the delineation result review and quality inspection in a timely and effective way, and jointly reviewed and determined the delineation result of "Three Zones & Three Lines" to ensure the authenticity, accuracy and usability of the results. The departments have reached consensus on the disposal of contradictions & conflicts, which has further improved the scientificity and rationality of the ERLs. The government departments at municipality (prefecture) and county levels have recognized the delineation results.

5.8.3 Benefits Earned from the Delineation

5.8.3.1 Set Up a Stable Territorial Ecological Security Barrier and Maintain the National Ecological Conservation Pattern

100% of glaciers and permanent snow cover, 88.12% of lake water surface, 65.56% of swamp and grassland, 64.14% of wetland, 59.47% of inland tidal flat and 54.23% of river water surface in Qinghai Province have been delineated within the ERLs, so that snow-capped mountains, glaciers, river headwaters, forest shrubs, grassland vegetation and desert vegetation in the province could be effectively protected.

The areas of Sanjiangyuan region and Qilian Shan region inside the ERLs have further increased, which could effectively conserve the biodiversity zones of global significance and the key national water conservation zones in Qinghai Province. The ERLs pattern composed of "one barrier, one shelterbelt and three ecological zones" has been formed, with one ecological barrier of Sanjiangyuan grassland, meadow and wetland, and one shelterbelt of Qilian Shan glacier and water conservation as the main body, with three ecological zones including Qinghai Hu grassland and wetland ecological zone, Qaidam desert and wetland ecological zone and eastern hilly biodiversity ecological zone as the key components. Such pattern has consolidated the Qinghai-Xizang Plateau ecological barrier, as well as optimized and improved territorial ecological security strategic pattern.

5.8.3.2 Significantly Improving the Integrity and Connectivity of National Parks, as Well as the Ecosystem Services

The area of Sanjiangyuan region and Qilian Shan region within the ERLs has further increased, which has further enhanced the ecosystem services of Sanjiangyuan and Qilian Shan national parks and realized the overall conservation of the ecosystem. It laid a foundation for building a national park demonstration province and an Ecological Civilization Demonstration Zone.

First, it connected the Chang Jiang source area and the Huang He source area in national parks (ancient Zonglie basin), thus built the wildlife migration channel in national parks and enhancing the ecological function of water conservation.

Second, it connected the source area of the Chang Jiang and the Qinghai Hoh Xil the World Natural Heritage (Tuotuo He area), which improved the integrity and connectivity of the ecosystem of Sanjiangyuan National Park.

Third, typical forests, shrubs and wetlands in Qilian Shan National Park have been added into the ERLs, which improved the ecosystem services function and quality of Qilian Shan National Park.

Fourth, freshwater lakes, lakeside wetlands, naked carp migration channels, and critically endangered Przewalski's gazelle habitats in Qinghai Hu area were delineated into the ERLs, which has laid a foundation for the further construction of Qinghai Hu National Park.

5.8.3.3 Fully Improving the Water Conservation Function and Effectively Maintaining the Ecological Value of Biodiversity

The extremely important ecological function areas, such as 83.31% of water conservation areas, 60.83% of windbreak and sand fixation areas, 60.16% of water and soil conservation areas and 94.15% of extremely fragile areas of soil erosion were delineated in the ERLs. 80.39% of Chang Jiang source, 53.77% of Huang He source, 62.62% of Lancang Jiang source, 49.83% of Hei He source, 76.59% of Shule He source and 99.78% of Shiyang He source were delineated into the ERLs, which could ensure the water conservation function of river sources, lakes and wetlands. It is of practical and far-reaching historical significance for continuously performing ecosystem service function, consolidating the role and status of "China's water tower", ensuring "clear stream flows eastward" and maintaining water ecological security in surrounding and downstream areas.

Meanwhile,89.00% of ecological functional areas for biodiversity conservation were delineated in the ERLs, which have focused on the conservation of the breeding habitats and migration channels for national and provincial key conserved species such as Tibetan antelope, snow leopard, black-necked crane, wild yak, Przewalski's gazelle, desert cat and naked carp, as well as the main rare plants and endangered wild plants such as Astragalus membranaceus, Gentiana Algida Pall, Gentiana futtereri, Gymnadenia orchidis Lindl, Kengyilia Thoroldiana and sand-fixing grasslands, etc.

It was in favor of conserving wildlife and their habitats in the Qinghai-Xizang Plateau, the preservation of rare and endangered species and the restoration of the wild animal population, so as to ensure the effective conservation of the richest and most integrated biological gene bank and the largest germplasm bank of wild species in the Qinghai-Xizang Plateau.

5.8.3.4 Properly Resolving Conflicts and Promoting Sustainable Economic and Social Development

Qinghai Provincial Government has carried out an evaluation and adjustment of the problems left over from history and the realistic conflicts in the ERLs, based on the actual situation and the need for future development, in a scientific, objective and factual manner. On the premise of not affecting the ecosystem services, the area of current land use and planned development space such as farmland, permanent basic farmland, mineral resources, water conservancy and hydropower, energy, tourism, key industries and poverty alleviation have reduced in a rational way, according to the national rules.

The government has properly dealt with the conflicts & contradictions among ERLs, poverty alleviation and major construction projects, thus resolving the conflicts among ecological conservation, economic and social development, which has laid a foundation for delineation and implementation of the three controlled lines without crossing, overlapping or conflict in territorial and spatial planning.

5.8.4 Thoughts on Further Management

At present, the ERLs of Qinghai Province have been delineated and incorporated into the "one map" of territorial and spatial planning for application and management. How to handle the relationship between conservation and development within

ERLs with an area of 296,500 km^2 in the future will be a major issue that needs to be studied and solved.

1. In view of the wide scope of the ERLs and the difficulty in supervision, it needs to improve the ecological compensation mechanism.

   The ERLs of Qinghai Province cover a wide scope and a large area, which have made great contributions to safeguarding the ecological security barrier in the middle and lower reaches of the Chang Jiang, the Huang He and even the whole country. However, Qinghai Province is a collection of the western region, plateau region, underdeveloped region, multi-ethnic region and key ecosystem service zones, with a low financial self-sufficiency rate and a large funding gap for ecological conservation, which cannot meet the needs of ecological conservation construction.

   It is suggested that, on the one hand, factors such as the high proportion of the ERLs and high cost of ecological conservation in Qinghai Province should be fully considered. The budget should increase greatly for key ecosystem service zones. On the other hand, an ecological compensation mechanism for grasslands, woodlands, wetlands, water resources and ecological resources within ERLs should be established. In addition, the compensation mechanism between river basins, regions and upstream and downstream should be established, and policy support should be given.
2. In view of the large number and large area of national parks within the ERLs, it needs to increase capital investment and policy support.

   It is suggested that, on the one hand, the central government should strengthen the project implementation and capital investment on ecological conservation and restoration within the ERLs with Sanjiangyuan and Qilian Shan national parks and protected areas as the main body, and also increase the investment in mining environment restoration and management, soil erosion control, grassland degradation control in Qinghai Province, so as to achieve the purpose of improving the ecosystem resilience and stabilizing the ecosystem services.

   On the other hand, it is suggested to innovate the ecological conservation system. Refer to the experience of national park pilots, the government explores the way of integrating ecological conservation, people's livelihood improvement and local economic and social development closely, so as to further promote the improvement of ecology.
3. In view of the scientific management of the ERLs, it is necessary to explore and coordinate the balance between development and ecological conservation continuously.

   The government needs to strengthen the practical exploration of ecological conservation and development and to seek a "win-win" path for conservation and development. On the one hand, the government should strictly implement the grass-livestock balance system within the ERLs, change the traditional production mode of grassland animal husbandry and reduce the grazing pressure of natural grassland through integrated measures, such as rotational grazing, grazing suspension and grazing prohibition, to realize the balance between grass and livestock and promote a virtuous cycle of grassland ecology.

On the other hand, the government should further strengthen the management and control of limited human activities in the ERLs, and implement the construction of a supervision information system with the help of "one map" of territorial and spatial planning, to improve the resource sharing with the departments of ecology and environment, forestry and grassland, thus forming a concerted supervisory approach.

5.9 Fujian Province: Conserving the Lucid Water and Lush Mountains[10]

Fujian Province, as the first national Ecological Civilization Pilot Zone in China, focuses on the demand for promoting high-quality development and transcendence in an all-round way, and has adhered to the principles of delineating into conservation area as much as possible. The government has given priority to conservation and taking development into consideration as well, evaluated the accuracy of basic data, the rationality of boundary demarcation, contradictions & conflicts, then adjusted and optimized the preliminary delineation. The provincial government fully communicated with local governments through the "up seven times and down seven times" process, namely, seven rounds of communication, then formed the results of the ERLs delineation.

The delineation work was carried out in accordance with the principles of overall planning of land and marine areas, coordinated management and delineation, overall stability and partial adjustment, and the three controlled lines without crossing, overlapping or conflicting in territorial and spatial planning. The main effects are as follows:

First, delineating into conservation areas as much as possible. In accordance with the national relevant regulations and rules, the areas of ecological space that provide important ecological function and must be strictly conserved, were brought within the scope of the ERLs framework. 70% of the ERLs area was concentrated in Wuyi Shan-Daiyun Shan, and the coastal shelter belts of key ecosystem function, which have built a solid provincial ecological security barrier.

Second, giving priority to conservation and taking development into consideration as well. Coordinating the planning of land, ocean, minerals, urban and rural areas, transportation, fishing ports, aquaculture, etc., Fujian Provincial Government adjusted the artificial commercial forests, approved construction

[10]This part was written by Shi Feihong, Lu Junjie, Wu Yifeng, Lin Wenyong and Peng Kai. Shi Feihong, an engineer of Fujian Land and Resources Survey and Planning Institute; Lu Junjie, the clerk of Fuzhou Natural Resources and Planning Bureau; Wu Yifeng, the director of Sanming Natural Resources Bureau; Lin Wenyong, the vice director of Longyan Natural Resources Bureau; Peng Kai, the section chief of Nanping Natural Resources Bureau.

land, sea island use rights, mining rights, etc., out of the scope of the ERLs. It has focused on ensuring key construction projects and people's livelihood projects at the provincial level, and reserved construction land space for promoting high-quality development in an all-round way.

Third, abiding by cultivated land conservation red line and rationally coordinating agricultural space. The reduction of cultivated land and permanent basic farmland has been avoided as far as possible.

Fourth, optimizing the boundaries, and improving the quality of the ERLs continuously. The preliminary red lines with the problems like the fragments, holes, serration boundaries, internal gaps etc., have been optimized. The boundaries of rivers and mountains have been checked and adjusted, so the results of the ERLs after adjustment were correct, which improved the connectivity and integrity of the ecosystem, so as to demarcate borders and calibrate of regions subsequently and implement the management. The area of territorial space designated for the ERLs was 43,400 km^2 in Fujian Province, building a solid ecological conservation pattern.

5.9.1 Fuzhou Municipality: Ecological Guarantee Barrier on Beach

Fuzhou features a unique natural ecology and is surrounded by mountains and lakes. Fuzhou has always adhered to the concept of prioritizing ecological conservation and green development, delineated the ERLs in a coordinated manner based on territorial and spatial planning, and carried out ecological conservation and restoration in an overall manner, having built a solid ecological security barrier for the seaside city.

5.9.1.1 Delineating ERLs in a Scientific Way

The total area within the ERLs is 5082.05 km^2 in Fuzhou Municipality. The main methods of delineating are as follows:

First, strengthening basic research. On the basis of systematic analysis of Fuzhou's natural geographical pattern and the results of the Third National Land Survey, ecosystem function zones were identified. The model evaluation method and quantitative index evaluation method were used to scientifically evaluate the key areas of water conservation, windbreak and sand fixation and biodiversity. The related department has conducted on-the-spot reconnaissance and investigation of the key areas to be included in the scope of ERL, and has verified the biological community, water quality, soil, etc., on the spot, which has improved the accuracy of the evaluation result and its consistency with the field.

Second, delineating ERLs as much as possible. 32 protected areas with a total area of 1008.07 km^2 after integration and optimization have been brought within the scope of the ERLs to realize ecological conservation as much as possible. Key territorial spaces, such as provincial/national ecological non-commercial forests, drinking water conservation areas at or above the county level, coastal backbone forests, mangroves and provincial key wetlands, were brought within the scope of ERL (Fig. 5.29).

Third, strengthening land and marine coordination. Using the coastline survey data as the boundaries between land and marine areas, carrying out the evaluation, adjustment and optimization of the land and sea respectively, the boundaries were spliced together. The topological errors such as ecosystem integrity, gaps and overlaps in the boundaries between the ERLs for land and marine areas were reviewed and processed, and finally the unified ERLs between land and marine areas were formed.

5.9.1.2 Building a High-Quality Ecological Pattern

On the basis of the delineation of the ERLs and the natural landscape pattern of Fuzhou, the integrated ecosystem conservation pattern composed of "one zone, four bays, five shelterbelts, five corridors and multiple wedges" has been constructed, and the corresponding guidelines for spatial planning have been put forward.

Fig. 5.29 Min Jiang Estuary Nature Reserve in Fujian

"One zone" is the western ecological conservation area. Its function is to strengthen ecological conservation in western mountainous areas, enhance forest ecological service function, and strengthen the natural restoration of species habitat (Fig. 5.30), so as to maintain biodiversity. It will improve the water conservation of river basins, and strengthen the water and soil conservation capacity of mountainous areas, thus conserving the sub-tropical native forest ecosystem comprehensively.

"Four bays" are four important bays, including Luoyuan Bay, Min Jiang Estuary, Fuqing Bay and Jiangyin Bay. The function is to strengthen the ecological conservation and restoration of key bay areas, through focusing on the conservation of key bays, islands and natural shorelines, thus enhancing the ecosystem services of coastal waters and coastal zones. These bays aim to conserve the ecology of coastal waters, and carry out restoration of wetland ecosystems in bays and estuaries such as mangroves, strengthen the conservation of migration routes and preferred habitats of rare waterfowl and marine wildlife.

"Five shelterbelts" are ecosystem service zones along five key rivers—Min Jiang (Wulong Jiang), Ao Jiang, Long Jiang, Dazhang Xi and Qibu Xi. The river system is connected in series with inland rivers, lakes, forests, wetlands, oceans and other ecosystems in the basin, so as to conserve the ecosystem service zones along key rivers, conserve and restore river ecosystems, strengthen the conservation of water conservation function in the upper reaches of rivers and the environmental management of river ecological corridors, and strengthen river course construction and ecological conservation.

"Five corridors" are five key mountain-sea ecological corridors—He Shan-Dabijia Shan, Changlong Jianfeng Shan-Baiyun Shan, Lianhua Shan-Gu Shan-Shoushi Shan-Min Jiang Estuary, Qi Shan-Dahua Shan-Huwei Shan-Nanyang Shan and Lingshi Shan-Fuqing Bay. The development and construction activities in the

Fig. 5.30 "Three Treasures" of Min Jiang Estuary Wetland-Chinese Crested Tern

ecological corridor are forbidden strictly. The width of the ecological corridors of mountain and sea is above 600 m.

"Multiple wedges" are green wedge zones between several cities. In line with the trend of mountains and the urban layout, it made plans for a number of green wedges between urban groups in the central city to avoid continuous groups and control the width above 200 m. The development and construction activities in green wedge zones are controlled strictly, so as to give full play to the ecological and leisure services of green wedge zones, and promote the connection between the urban ecosystem and other areas.

5.9.1.3 Strengthening Ecological Space Governance

To carry out integrated conservation and systematic management in the Min Jiang Basin, the government of Fuzhou municipality adhered to "land and sea coordination and river and sea linkage", and formed a "mountain-sea" model of ecological conservation and restoration in Fuzhou.

First, strengthening the top-level design. Guided by the overall planning of territorial space, the authority integrated ecological conservation and restoration with landscape architecture, and land consolidation etc., and carried out systematic restoration. It refined the restoration objectives and planning strategies to specific project lists and formulated implementation plans to promote the implementation of objectives.

Second, strengthening technical support. The government of Fuzhou municipality constructed the technical support mode of "local city-trans cities-trans regional", explored the mode of "tech correspondent" in the field of ecological conservation and restoration and accelerated the transformation of scientific research achievements in ecological conservation and restoration.

Third, insisting on the promotion of projects. A total of 89 restoration projects have been planned for the entire basin of the Min Jiang, 1.7 billion yuan of project funds has been raised, of which 203 million yuan has been supported by the central government. At present, three excellent demonstration projects have been formed, which provide a model for ecological conservation and restoration in river basins.

5.9.2 Longyan City: Green Revolution

5.9.2.1 Building an Ecological Security Pattern and a Solid Ecological Security Barrier

A total area of 6390.08 km^2 has been designated into ERLs in Longyan City, including national protected areas, primary and secondary conservation areas of drinking water sources, scenic spots, national geological parks, etc., realizing the goal of important ecological space areas with important ecological functions, extremely fragile ecology areas are controlled by the red lines. Longyan City constructed an ecological conservation configuration composed of "three barriers, three corridors and multiple nodes".

"Three barriers" serve as the ecological barriers with the Wuyi Mountains, Daimao Mountains and Bopingling Mountains. The main function is to conserve the natural mountain ecosystem, strengthen the regional ecological security, ensure the functions of water conservation, soil erosion control and biodiversity conservation, and build a solid ecological security barrier.

"Three corridors" are the three water systems—Ting Jiang, Jiulong Jiang and Min Jiang. The main function includes connecting through ecological corridors, connecting key protected areas and species habitat in forests, and linking the ecological conservation spaces such as rivers, lakes and wetlands.

"Multiple nodes" refer to three protected areas, 10 forest parks, three scenic spots, three wetland parks, and other 214 spots (the first batch of key protected wetland zones in Fujian), such as protected areas, Mianhuatan Reservoir and Zhangping Nanyang Wetland. These patches have been brought within the scope of the ERL framework, so as to improve the stability and ecological service of natural ecosystems.

5.9.2.2 Sustainable Ecological Conservation

On the basis of the ERLs as a lifeline of territorial ecological security, Longyan City has continued to explore new paths of green development, and ecological practice models like "Changting experience" and "Wuping experience" that can be replicated and promoted throughout the country have emerged, thus becoming the vanguard and pioneer in practicing thought on Ecological Civilization. The main results are as follows:

1. Forestry ecology continues to lead the way, and forest ecological conservation has achieved remarkable results. Longyan is a national forest city, a national greening model city, a national comprehensive pilot city for forestry reform and development, and the first national forestry science and technology demonstration area in Fujian Province, with a forest coverage rate of 79.21% and a forest volume of 157 million m^3. The "Wuping Experience" of forest reform has achieved remarkable results, and the comprehensive pilot construction of forest reform development has been promoted by establishing mechanisms, such as the transfer of forest land management rights.
2. Propagating the "Changting Experience" continuously and promoting the high-quality protection of soil erosion. Longyan was once a region of severe soil erosion and water loss, especially Changting County was one of the most severe soil erosion and water loss in the red soil region of southern China before. Through unremitting efforts by the government, remarkable achievements have been made in soil erosion control (Fig. 5.31). In 2020, the area of soil and water loss

Fig. 5.31 Ecological governance of Ting Jiang Basin in Longyan

in the whole city reduced to 1147.40 km^2, and the rate of soil and water loss reduced to 6.03%, among which the area of soil and water loss in Changting County reduced to 210.13 km^2, and the rate of soil and water loss reduced to 6.78%. The practice of comprehensive control of soil and water loss and ecological restoration in Changting County was selected as a typical case of ecological restoration at the 15th Conference of the Parties to the UN Convention on Biological Diversity (COP15). "Changting Experience" has been known as a model of soil erosion control in China, and has become a banner of ecological construction in Fujian, which has been affirmed by President Xi.

3. Ecological disadvantages are complemented, and water environment quality has steadily improved. The government of Longyan City has promoted the water environment treatment and ecological restoration projects of key wetlands and drinking water sources within the scope of the ERLs, and significantly improved the ecological quality and ecological functions of the basins. The ecological compensation mechanism of the Tingjiang-Hanjiang inter-provincial basin has been put on the list of reform measures and experience promotion in the national Ecological Civilization Pilot Zone.

5.9.2.3 Promoting Green Development

The wetland parks, forest parks and geoparks were taken as carriers according to the plan. The government of Longyan City carried out the construction of wetland ecological scenic spots and ecological green corridors, implemented water conservation projects, forest conservation programs and ecological landscape forest upgrading projects, created Longyan World Geopark, built a 210 km forest trail branch line with seven counties involved, and created the characteristic travel resources with the combination of Longyan history and the environment. In 2020, forest scenic spots in Longyan received 11.84 million tourists, bringing in direct revenue of 1.073 billion yuan, with a total social output value of 4.738 billion yuan. By clarifying the property rights of ecological resources and speeding up ecological industrialization and industrial ecologicalization, Longyan has created the ecological economic brand system of "alpine economy and under-forest economy", bringing vitality to green development.

Taking ecological construction as the foundation, Longyan has conserved and utilized ecological resources in an effective and rational way through the delineation of the ERLs, which promoted the green and low-carbon development of society and economy, and the benign interaction between ecology and industry.

5.9.3 *Sanming City: Mountain Belt*

Sanming City, located in the west of Fujian Province, is praised as the green capital of China, national forest city, and the national Ecological Civilization Demonstration Zone. Sanming has three world-class and more than 200 national tourism brands, such as Taining World Natural Heritage Site and World Geopark. The total area of 6479.20 km^2 has been delineated into the ERLs in Sanming City, accounting for 28.21% of the municipality area, and occupied 1/5 of the ERLs of land area in Fujian Province. The conservation area ranked second in Fujian Province. Based on the delineation of ERL, Sanming has built the ecological security pattern of "three mountains, three lakes, one core and multiple cores", connecting Nanping and Longyan to jointly build a green development zone in mountainous areas of Fujian Province.

5.9.3.1 Giving Priority to Ecological Conservation and Delineating ERLs as Much as Possible

Guided by maintaining ecological function and ensuring ecological security, the Sanming government has taken into account the integrity and systematicness of natural ecology in a coordinated manner, and brought the integrated and optimized protected areas, key wetlands and extremely important ecological function areas as well as the areas that need to be strictly conserved in the city within the scope of the ERL framework, on the basis of evaluating the importance of ecological function. For example, Taining World Natural Heritage Site, covering an area of 243.65 km^2, known as the only representative of the youth "Mountain Plateau-Canyon" Danxia in China's sub-tropical humid areas and the place where China's Danxia story begins, is praised by international geologists. Another example is the Zhengyuan Zone of Min Jiang basin, which is a key part of mid-sub-tropical forest ecosystem reserves in the Wuyi Shan, a preferred habitat for rare wildlife and gene bank for precious species. The total area, 150.59 km^2, has been brought within the scope of the ERLs framework, thus creating the ecological core area of the Min Jiang source-Dajin Hu region, and fully implementing the spaces for ecological conservation in an overall and systematic manner.

5.9.3.2 Adhering to Seeking Truth from Facts and Proceeding from the Reality of Development

Sanming City comprehensively sorted out the fragmented in the images within the scope of former ERLs, identified the contradictions & conflicts between ERLs and the boundaries of permanent basic farmland, cultivated land, towns and villages, approved construction land, major infrastructure projects and major construction projects above the provincial level, legal mining rights, artificial commercial forests, etc. Then it carried out evaluation and adjustment in a coordinated manner, to meet the rational development demand and the legitimate rights and interests of the parties concerned. The speckle patches have been optimized dropped to 7419 from 14,600. The area involved in conflicts that have been resolved is about 660 km^2. For the delineated ERLs in a coordinated manner, Sanming ensured that the ecological services were maintained, the area of ERL did not decline, and the nature did not change. The effective ecological conservation promoted high-quality development.

5.9.3.3 Strengthening Work Coordination and Promoting Results Sharing

Sanming took the delineation of the ERLs and strictly adhered to it as key tasks of developing the territorial and spatial planning and ecological conservation work, and it was included in the annual key tasks of the government of Sanming City. The government has set up a leader group in charge of preparing the overall territorial and spatial planning for Sanming, established a coordination mechanism for planning preparation, set up a special technical group for the delineation of the ERLs, clarified the division of responsibilities among departments, strengthened work communication and coordination, and promoted the delineation of the ERLs in a coordinated manner. Since the results of ERL delineation took effect, Sanming City has strictly adhered to them, and has been prohibited from carrying out development activities that did not align with ecological services. The natural resources authorities at city and county levels participated in the site selection work in the early stage of the project implementation, conducted a review of the project site selection in strict accordance with the ERLs control rules, the Bureau of Natural Resources, Sanming City, issued a document to further emphasize the implementation of the ERLs and other control requirements in the early stage of project site selection and route selection.

5.9.4 *Nanping City: Wuyishan National Park*

Nanping City is the source of the Min Jiang. Wuyi Shan in Nanping is the boundary of the Southeast coastal hills and Jiangnan hills, both of which are treasure houses of biodiversity and key ecological security barriers in the East of China (Fig. 5.32). Nanping took the responsibility for ecological security barrier in Fujian Province. The ERLs in the entire city covered 7551.13 km^2, accounting for 28.73% of Nanping City and 23.89% of the land ERLs area in Fujian, and the total area delineated into the ERLs ranked first in Fujian Province.

5.9.4.1 Shelterbelt Construction, Created a Pilot Demonstration Zone for Harmonious Symbiosis Between Human and Nature

The government of Nanping City gave priority to ecological conservation, focused on conservation and development in a coordinated manner, continued to deepen the innovation of system and mechanisms, comprehensively promoted the construction of a protected area system with Wuyishan National Park as the main body, and explored the replicable and shareable experience of national park conservation. Nanping has explored a model of national park shelterbelt construction for conservation and development with a more comprehensive, more macro angle. Through in-depth surveying and learning from advanced experience and practices, Nanping City has designated an area of 4252 km^2, where it is 1001.41 km^2 away from

Fig. 5.32 Jiuqu He at Wuyi Shan scenic area

Wuyishan National Park, as the shelterbelt for conservation and development of Wuyishan National Park, that is, the "outer ring" shelterbelt was used to conserve the "inner ring" national park, and Nanping explored to build a pilot demonstration zone for harmonious symbiosis between human and nature. By strengthening the ecological conservation of the "outer ring", the city has further enhanced the authenticity and integrity of the natural ecosystem of the national park, conserved biodiversity and ensured the ecological security barrier.

5.9.4.2 Demonstration Zone for Ecological Civilization Construction

Taking the "shelterbelt" construction as the key measure, the government of Nanping City demonstrated and took the lead in the construction and conservation of national parks. It strengthened the construction and restoration, continued to strengthen the integrated application of monitoring and management, such as remote sensing technology, space-air-ground integrated technology and big data analysis, enhanced the stability of ecological barrier in the "shelterbelt" region. It encouraged relevant functional departments, well-known universities and scientific research institutions to set up research institutes (bases) in Nanping, such as biodiversity conservation, wildlife conservation, and ecological scientific research, to provide talent and intellectual support for the conservation of national parks. It strengthened legislative conservation, promoted the unity of ecological legislation, judicature and law enforcement, and conserved the ecological environment with the strictest system and the strictest rule of law.

5.9.4.3 Exploring the Way to Realize the Value of Ecological Products

Focusing on the requirements of "green development", the government regarded the national park as the major ecological product, and "shelterbelt" as the practice base for the transformation of "two mountains". It made the cultural and tourism economy better and stronger, actively promoted the construction of "No.1 Scenic Road of National Park", through integrating the scattered natural scenery, pastoral landscape, traditional culture and cultural landscape with the resources, thus striving to create a number of cultural tourism projects that fully display the characteristics of the natural ecosystem, natural landscape, cultural heritage and biodiversity, and promoting "nice scenery" to become a "new experience". The city government optimized the green economy, made good use of national parks and double-world heritage brands, deepened the exploration of the path to realize the value of ecological products, and accelerated the extension and supplement of the chains of ecologically advantageous industries such as tea (Fig. 5.33), bamboo, water and cultural tourism, thus promoting integrated development of the primary, secondary and tertiary industries to seek benefits. It built a "green industry innovation platform", thus promoting the integration of the "four chains" (industrial chain, talent chain, innovation chain and capital chain). It boosted the transformation of "two mountains"

Fig. 5.33 Swallow Nest ecological tea garden

through technology, so as to turn ecological advantages into competitive advantages and beautiful scenery into a beautiful economy.

5.9.4.4 Building a "Happiness Circle" That Is Livable, Suitable for Business and Travel

The government of Nanping City has always adhered to the goal of "improving people's livelihood", insisted that a good ecology is a great welfare of people's livelihood, emphasized the cultural benefits, ecological benefits, and green prosperity for the people, and accelerated the construction of national park to achieve common prosperity. On the one hand, it gave full play to the radiation leading role of national parks, took industrial revitalization as the starting point, and speeded up the creation of a number of rural revitalization demonstration sites featuring flexible mechanism, excellent industry, rich people and beautiful ecology. On the other hand, it speeded up the construction of the gateway towns surrounding national parks, focused on promoting a new type of people-centered urbanization, made efforts to fill shortcomings in infrastructure, and cultivated a convenient living circle with multi-format agglomeration and a service circle with multi-functional service agglomeration, so that everyone can enjoy a quality life, take root and grow, and have a stage to show themselves.

5.9.5 *Suggestions on Strengthening the Control of the ERLs*

5.9.5.1 Clarifying the Responsible Subject and Strengthening Cooperative Conservation

Local governments at all levels should take overall responsibility for the delineation, evaluation, adjustment and management of the ERLs within their respective administrative areas, and the departments of natural resources, ecology, forestry, agriculture and rural areas perform their management and control duties according to their respective functions, thus forming synergies in ecological conservation.

5.9.5.2 Improving the Control System and Strictly Implementing Management

Relying on the basic information platform for territorial space, the ERLs should be integrated into the "one map" territorial and spatial planning as the basis for territorial space use control. By taking measures such as usage permits, development intensity and utilization methods, the government can ensure that human activities within the scope of the ERLs are controlled within the approved range. Through the integrated management and control of natural and human elements, the province will coordinate the relationship between ecological conservation, food security and economic development.

5.9.5.3 Carrying out Marking on Site to Realize Fine Management

Arranging boundary surveying on site, calibration and mapping of the ERLs to ensure the accurate implementation. The government should clarify the boundary direction of the red lines block and the coordinates of inflection points on the spot, conduct reconnaissance of the boundaries of the ERLs, determine the accurate spatial coordinates of boundary markers, set up a database of boundary markers, and formulate the map. By means of setting up electronic boundary piles and electronic fences, the government should set up unified and standardized entity boundary piles and identification signs at key control points, such as key sections and inflection points. No unit or individual may destroy, deface or remove ERL boundary markers or signs without authorization.

5.9.5.4 Strengthening Treatment and Restoration to Maintain Ecological Functions

It is supposed to take the county-level administrative region as the basic unit, according to the regional leading ecological function and ecosystem structure, the plans for ecosystem conservation and restoration should be formulated and implemented. Giving priority to conserving good ecosystems and preferred habitats of key species, the government should establish and improve ecological corridors, and improve the integrity and connectivity of ecosystems. In areas with suitable conditions, ecological migration can be gradually promoted, moderate centralized resettlement of the population can be promoted in an orderly manner. Thus, the intensity of human activities and ecological pressure can be reduced. In accordance with the principle of overall planning and management of land and marine, the government should carry out ecological governance and remediation of the ERLs of marine, prohibit the implementation of artificial reef throwing, strengthen the conservation of the ecological environment and prevention and treatment of pollution in the ERLs and surrounding areas, and focus on strengthening the integrated governance of the rivers entering the sea within the scope of ERLs.

5.9.5.5 Improving the Mechanism Construction and Promoting Joint Conservation and Governance

First, improving the diversified ecological compensation mechanism. According to the principle that one who receives benefit should give compensation, it is necessary to establish and improve the market-oriented and diversified ecological compensation mechanism by implementing emission rights, carbon emissions, water rights and forest rights. The government should guide ecological beneficiaries to compensate ecological protectors, as well as arouse the enthusiasm of local governments to conduct supervision through incentive measures, thus improving the transfer payment system in national key ecosystem service zones and promoting the implementation of ERL rules.

Second, establishing mechanisms for realizing the value of ecological products. It is necessary to promote the value transformation of ecological products, through the combination of government and the market, thus combining ecological conservation and improving people's livelihood.

Third, improving the public participation mechanism. It is necessary to increase the delineation of ERL and the disclosure of monitoring information, and guide the public to participate in the evaluation, adjustment and supervision of ERL, thus giving full play to the subjective initiative of the public who have the rights and responsibilities for ecological conservation.

Correction to: The Blue Book on China's Ecological Red Lines 2023

Correction to:
China Land Surveying and Planning Institute,
***The Blue Book on China's Ecological Red Lines* 2023,**
https://doi.org/10.1007/978-981-95-0051-2

The copyright holder of the chapters and back matter have been retrospectively corrected. The correct copyright holder name is © China Land Press Ltd. 2025.

The updated version of this book can be found at
https://doi.org/10.1007/978-981-95-0051-2

China Land Surveying and Planning Institute, *The Blue Book on China's Ecological Red Lines 2023*, https://doi.org/10.1007/978-981-95-0051-2_6

Appendix: Glossary

1. Ecological Red Lines (ERLs)

   The ERLs are the lifelines of territorial ecological security. China has brought extremely important ecological function areas, extremely fragile ecological areas and areas with potentially important ecological value within the scope of the ERL framework.
2. Ecological space

   It is a functional space that ensures ecological security, provides ecosystem services or ecological products, and mainly ecological land.
3. System of protected areas

   It refers to a territory or marine area with important natural ecosystems, natural relics, natural landscapes, ecological functions or cultural values, designated or confirmed by governments at all levels, and carried out for long-term protection. According to the authenticity, integrity, systematization and internal laws of the natural ecosystem, according to the management objectives and efficiency and international experience, the protected natural areas are divided into national parks, nature reserves and nature parks according to their ecological value and protection intensity.
4. Biodiversity

   It is composed of the ecological complex organisms (animals, plants, microorganisms) and the environment, and various related ecological processes, including ecosystem diversity, species diversity and gene diversity.
5. Resources and environment carrying capacity and territorial development suitability

   Resources and environment carrying capacity refers to the maximum and reasonable volume of natural resource elements within a particular area that can support human activities such as agricultural production and urban construction. It is based on a specific stage of development, the level of economy and technology, the way of production and life and ecological conservation.

China Land Surveying and Planning Institute, *The Blue Book on China's Ecological Red Lines 2023*, https://doi.org/10.1007/978-981-95-0051-2

Territorial development suitability refers to the appropriate degree of human activities, such as agricultural production and urban construction in a specific land space.

6. Uninhabited island

   It refers to an island that does not have a registered residential address of residents.

References

Han Z Q, Zhang Y L, An X L et al (2023) Research progress on seagrass meadow restoration and eels transplanting cultivation techniques. Hebei Fisheries (2):32–36. (in Chinese with English abstract)

Li Y L, Gao Y, Zheng X Q et al (2022) Mechanism and evaluation method of coral reef ecosystem function. Journal Applied Oceanography 41(3):475482. (in Chinese with English abstract)

Li J R, Qiao L, Liu W L et al (2023) Study on ecological red lines, delineation and control of uninhabited Islands. Natural Resources Information. http://kns.cnki.net/kcms/detail/10.1798.N.20230301.1345.004.html. (in Chinese with English abstract)

Liu S L (2020) Optimize the ecological conservation pattern and enhance regional ecosystem services. https://m.gmw.cn/baijia/2020-12/02/34421383.html. (in Chinese with English abstract)

Xu B X, Zhang J, Lang S K et al (2022) Present situation and degradation factors of seaweed meadow along the coast of Wenchang, Hainan. Journal Applied Oceanography 41(4):614624. (in Chinese with English abstract)

Yang B J, Gao J X, Zou C X (2014) Strategic significance of delineating ecological red lines. China Development 14(1):4. (in Chinese with English abstract)

Zhang C, Yao X J, Xiao J S et al (2023a) Evolution characteristics of glaciers and lakes in Qinghai Province between 2000 and 2020. Journal of Natural Resources 38(3):822838. (in Chinese with English abstract)

Zhang X, Chen J Y, Yang Q J (2023b) Evolution of temporal and spatial distribution of mangroves in the greater bay area and remote sensing analysis of existing forest age. Acta Oceanologica Sinica 45(3):113124. (in Chinese with English abstract)

Zou C X, Peng H F, Liu C Y (2021) Thoughts on ensuring territorial ecological security in the new era. Environmental Protection 49(22):50–53. (in Chinese with English abstract)

China Land Surveying and Planning Institute, *The Blue Book on China's Ecological Red Lines 2023*, https://doi.org/10.1007/978-981-95-0051-2

MIX
Papier aus verantwortungsvollen Quellen
Paper from responsible sources
FSC® C105338

FSC
www.fsc.org

If you have any concerns about our products,
you can contact us on
ProductSafety@springernature.com

In case Publisher is established outside the EU,
the EU authorized representative is:
Springer Nature Customer Service Center GmbH
Europaplatz 3, 69115 Heidelberg, Germany

Printed by Libri Plureos GmbH
in Hamburg, Germany